I0041124

Window Design Strategies to Conserve Energy

S. Robert Hastings
Richard W. Crenshaw

Architectural Research Section
Center for Building Technology
Institute for Applied Technology
National Bureau of Standards
Washington, D.C. 20234

Sponsored by:

Energy Research and Development Administration
Washington, D.C. 20001

U.S. Department of Housing and Urban Development
Washington, D.C. 20410

Fredonia Books
Amsterdam, The Netherlands

Window Design Strategies to Conserve Energy

by
Center for Building Technology
National Bureau of Standards
U.S. Department of Urban and Housing Development

ISBN: 1-4101-0863-5

Copyright © 2005 by Fredonia Books

Reprinted from the 1977 edition

Fredonia Books
Amsterdam, The Netherlands
http://www.fredoniabooks.com

All rights reserved, including the right to reproduce
this book, or portions thereof, in any form.

TABLE OF CONTENTS

ABSTRACT

A multitude of design strategies are available to achieve energy-efficient windows. Opportunities for improving window performance fall into six groups: site, exterior appendages, frame, glazing, interior accessories, and building interior. Design strategies within these groups can improve one or more of the six energy functions of windows: solar heating, daylighting, shading, insulation, air tightness, and ventilation. Included in this report are 33 strategies; an explanation of the physical phenomena responsible for each strategy's energy performance, summarized energy and non-energy advantages and disadvantages; aesthetic considerations; cost approximations; example installations, laboratory studies, or calculations by the authors; and references. Intended readers include professional designers, lessees and owners of commercial space, home buyers and owners, window component manufacturers, and researchers. The report's purpose is to draw attention to the wide range of options currently available to conserve energy with windows

Key Words: Air-tightness; daylighting; energy conservation; insulation; shading; solar heating; ventilation; windows.

INTRODUCTION

Windows can substantially alter the amount of purchased energy
required to maintain comfort. Well-designed, they can actually
provide a net energy gain; poorly designed, they can be an enormous
energy burden. This report provides design strategies to make
windows energy conserving. Each strategy is directed at improving
one or more of the six energy functions of windows, which are:
providing winter solar heat, providing year-round daylighting,
rejecting summer solar heat, providing insulation and air tightness
during periods of heating or air conditioning, and providing natural ventilation during temperate weather.

Window design strategies to conserve energy need not be limited to
the frame and glazing. Site strategies can minimize adverse climatic forces and/or amplify beneficial climatic forces; exterior
appendages and interior accessories can supplement the capabilities
of the window frame and glazing; and building interior strategies
can insure maximum benefit is derived from the energy assets the
window provides.

In order to facilitate successful use of design strategies, this
report includes a cursory explanation of the physical phenomena
responsible for each stategy's performance. This is followed by a
list summarizing the energy, as well as non-energy, advantages and
disadvantages of each strategy. Then, since windows and their
accessories can drastically affect the quality of the building
exterior, as well as the character of the building exterior, aesthetics are discussed. Brief price inquiries are reported to
provide estimates of first costs. Installation was not included in
most cost figures because of the wide variation installation situa-
tions introduce. More precise dollar figures should be obtained
from local distributors before life-cycle costing is calculated for
an actual building. Finally, the references used in writing and
illustrating each strategy, and sources for further information,
are listed.

Selection of individual strategies should be based on the importance
of each of the window's energy functions, considering the local
climate, the time of day and/or seasons the building is most used,
and the environmental requirements of the activities being housed.
The strategy/function cross-reference table following the introduction is provided to help select strategies addressing the energy
functions determined to be most important for a specific project.
Final evaluation requires recalculation of the total window system,
since the performance of strategies in combination may differ from
the sum of each individual's performance.

The report is not only intended for the professional designer, but also for the researcher, to suggest further investigation of the many energy-conservation potentials of windows; the manufacturer, to encourage further refinement of the energy-conservation qualities of his product(s); and the commercial lessee or home buyer, with the hope that more energy-efficient windows will result as a consequence of demand from a consuming public better informed of the range of energy-conserving options available.

Work on this report has been conducted within the framework of an National Bureau of Standards interdisciplinary research project on the energy-related performance of windows. The work on this project was partially in support of the development of Building Energy Performance Standards. It was jointly supported by the Energy Research and Development Administration (Contract E(49-1) 3800), by the United States Department of Housing and Urban Development (Contract No. RT 193-12), and by the National Bureau of Standards. The content of the report draws heavily upon research and data published by private industry. In referencing this material, NBS in no way endorses specific manufacturers or products. Professional judgment must be exercised in assessing the capabilities of strategies singly or in combination for specific building projects.

STRATEGY/Function CROSS-REFERENCE TABLE

	Solar Heating	Daylighting	Shading	Insulation	Air Tightness	Ventilation
1. SITE						
WINDBREAKS				X	X	X
SHADE TREES			X	X		
GROUND SURFACES	X	X				
ORIENTATION TO SUN	X		X			
ORIENTATION TO WIND						X
2. EXTERIOR APPENDAGES						
SUN SCREENS			X	X		
EXTERIOR ROLL BLINDS			X	X		
ARCHI. PROJECTIONS		X	X			
EXTERIOR SHUTTERS			X	X	X	
AWNINGS			X			
3. FRAME						
FRAME VENTILATORS						X
WEATHERSTRIPPING					X	
THERMAL BREAK				X		
TYPE OF OPERATION					X	X
WINDOW TILT	X					
SIZE, ASPECT RATIO					X	
4. GLAZING						
MULTIPLE GLAZING				X		
HEAT-ABSORBING	X		X			
REFLECTIVE GLASS			X			
APPLIED FILMS			X	X		
REDUCED GLAZING			X	X		
GLASS BLOCK	X	X		X		
THRU-GLASS VENTILATORS						X
5. INTERIOR ACCESSORIES						
VENETIAN BLINDS		X	X			
DRAPERIES			X	X		
FILM SHADES	X		X	X		
OPAQUE ROLL SHADES	X		X			
INSULATING SHUTTERS				X		
6. BUILDING INTERIOR						
FIXTURE CIRCUITING		X				
TASK LIGHTING		X				
AUTOMATIC SWITCHING		X				
INTERIOR COLORS		X				
THERMAL MASS	X					

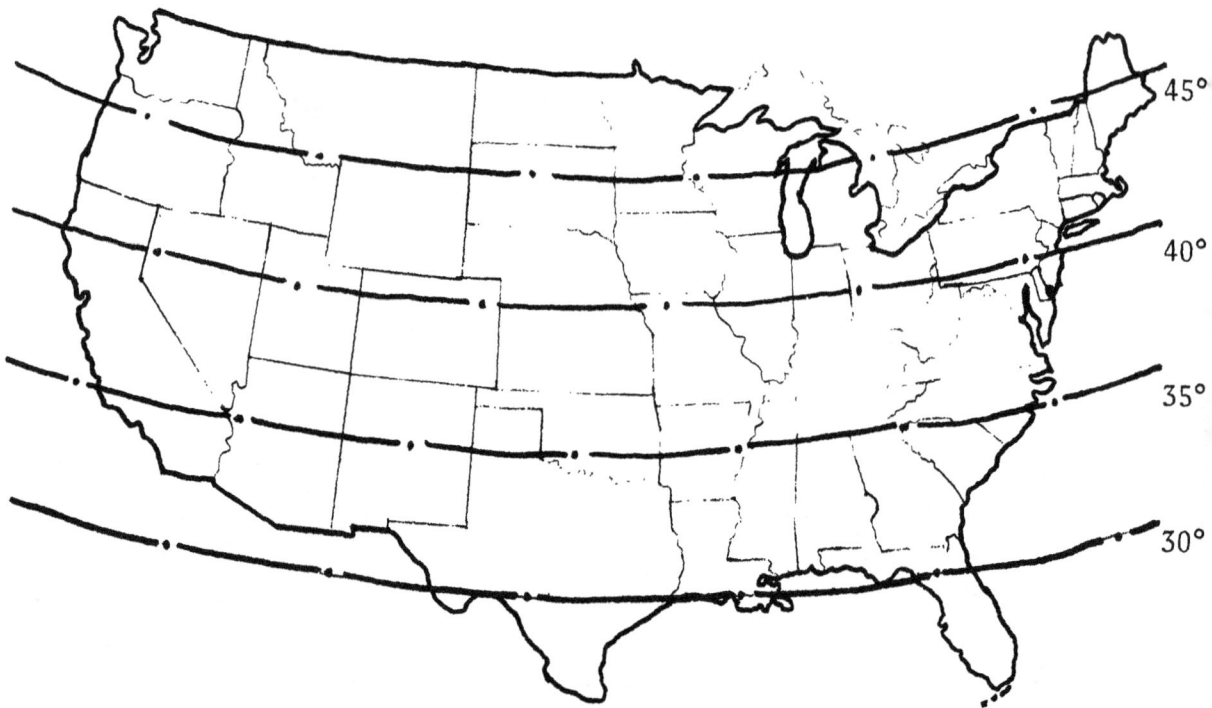

LATITUDES OF THE UNITED STATES

GLOSSARY

SOLAR HEATING:

Total Solar Transmittance = the amount of solar energy transmitted through the window configuration divided by the
total energy incident (striking) on the outside surface.
(Given as a percentage.)

Latitude: for the convenience of the reader a map of the U. S.
with latitudes is provided on the opposite page.

DAYLIGHTING:

Visible Transmittance = the amount of visible light transmitted
through a window configuration divided by the amount of
visible light incident on the outside surface. (Given
as a percentage.)

SHADING:

Shading Coefficient = the total amount of heat transmitted by
a window configuration divided by the total amount of heat
transmitted by a single pane of double strength (1/8")
glass.

INSULATION:

Btu = amount of heat required to raise the temperature of one
pound of water at its maximum density one degree Farenheit.

U-value = the amount of heat conducted from the inside air,
through the window configuration to the outside air, or
vice versa, in the summer. A 15-mph outside wind is assumed
in the winter, a 7 1/2-mph outside wind is assumed in the
summer. The air inside is assumed still summer and winter.
(Given in Btu/square foot, hour degree Farenheit difference
between outside and inside temperatures.) For example, a
U-value of 0.50 means 0.50 Btu's pass through each square
foot of window configuration for every hour and, for each
Farenheit degree difference existing between the inside
and outside temperature.

R-value = the resistance a material offers to the flow of heat
from one surface to another. The resistance between the
surface to the air is considered separately. The reciprocal
of the Sum of the Resistances equals the U-value. $U=1/(R_1+R_2+...R_n)$.

AIR TIGHTNESS, VENTILATION:

Volume of air passing through an opening = cubic feet of air
for a given time period divided by crack length (for
infiltration) or divided by open area of sash (for
ventilation).

SI CONVERSION

In view of present accepted practice in this country in this technological area, common U.S. units of measurement have been used throughout this paper. In recognition of the position of the USA as a signatory to the General Conference on Weights and Measures, which gave official status to the metric SI system of units in 1960, we assist readers interested in making use of the coherent system of SI units by giving conversion factors applicable to U.S. units used in this paper.

Length

1 in = 0.0254 meter* (m)
1 ft = 0.3048 meter* (m)

Area

1 ft^2 = 0.0929 square meter (m^2)

Volume

1 ft^3 = 0.0283 cubic meter (m^3)

Mass

1 lb = 0.453 kilogram (kg)

Mass/Volume (Density)

1 lb/ft^3 = 16.02 kilogram/meter^3 (kg/m^3)

Temperature

degree Celsius (°C) = 5/9 (°F - 32)

Volume/Time (Flow)

1 cfm = 0.000472 meter^3/second (m^3/s)

Velocity

1 mph = 0.447 meter/second (m/s)

Quantity of Heat

1 Btu = 1055.87 joule (J)

Thermal Resistance

1 °F h ft^2/Btu = 0.176 square meter degree Celsius/Watt (m^2.°C/W)

* exactly

SITE

WINDBREAKS
SHADE TREES
GROUND SURFACES
ORIENTATION TO SUN
ORIENTATION TO WIND

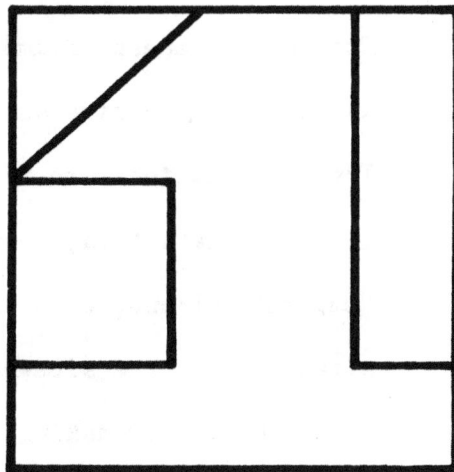

1. SITE

Two key site considerations are the siting of a building
to provide favorable window orientation to the sun and
favorable window orientation to prevailing winds. Landscaping can also improve window performance. Shade trees
can seasonally control direct radiation from the sun;
ground surfaces can control reflected radiation; planted
ground cover can moderate air temperature; and windbreaks
can diminish the force of the wind.

In general, there are two advantages to improving window
performance through site design. First, adverse climatic
forces are moderated at a distance. Residual forces are
then dissipated before encountering the windows. Second,
climatic moderation is likely to improve the performance
of adjacent walls and roofs, as well as windows.

The effectiveness of improving the energy performance of
windows through site design is apparent if heat-gain/heat-
loss calculations are first computed with unadjusted regional
climatic statistics, then recalculated considering the
tempering of sun, wind and air temperature possible through
site design. The site-moderated window performance should
be the basis for making trade-offs in window design.

1.1   WINDBREAKS / Air Tightness, Insulation, Ventilation

STRATEGY:

 Install a fence and/or row of trees or shrubs as a wind barrier
 to reduce wind pressure on windows.

PHENOMENA:

1) Air infiltration through windows can be reduced by di-
   minishing wind pressure by means of a windbreak. The most
   effective location for a windbreak is upwind a distance of
   1 1/2 to 2 1/2 times the height of the building. At this
   distance, the wind will be deflected up and well over the
   building, reducing the pushing action on the building's
   windward side and the pulling action on its leeward side.

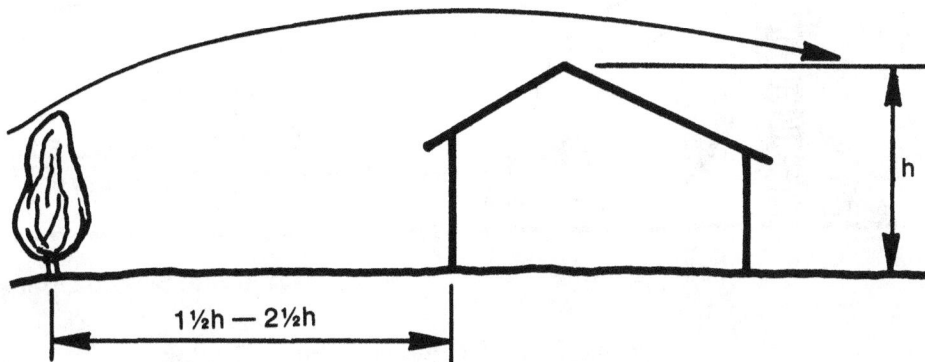

Figure 1.   Windbreak Distance

2) A windbreak is more effective if it allows part of the wind
to penetrate. A solid windbreak creates a low pressure area
on its leeward side with resulting strong eddy currents.
These may be as destructive as direct wind in eroding the
still air film at the surface of the window. Allowing a
portion of the wind to pass through the windbreak tends to
relieve this leeward suction. This is illustrated in the
following comparison between a solid wall and an open fence.

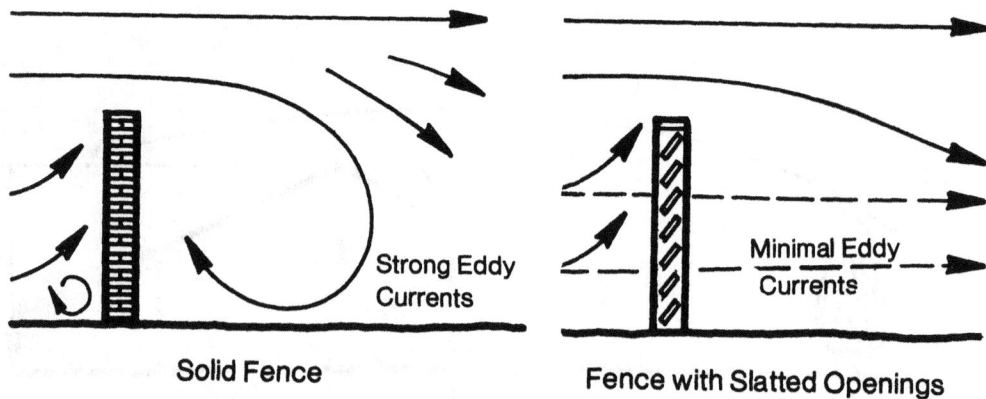

| Solid Fence | Fence with Slatted Openings |

Figure 2. Airflow vs. Fence Design

3) Heat transmission through windows can be reduced by di-
    minishing the amount of wind flowing across the glass.
    Glass is a good conductor of heat and therefore affords
    little impediment to heat flow. However, a still layer of
    air at the surface of the glass does retard heat flow.
    Blocking the wind will protect this boundary layer of air
    from the scouring force of wind. The importance of the
    boundary layer of air is illustrated below:

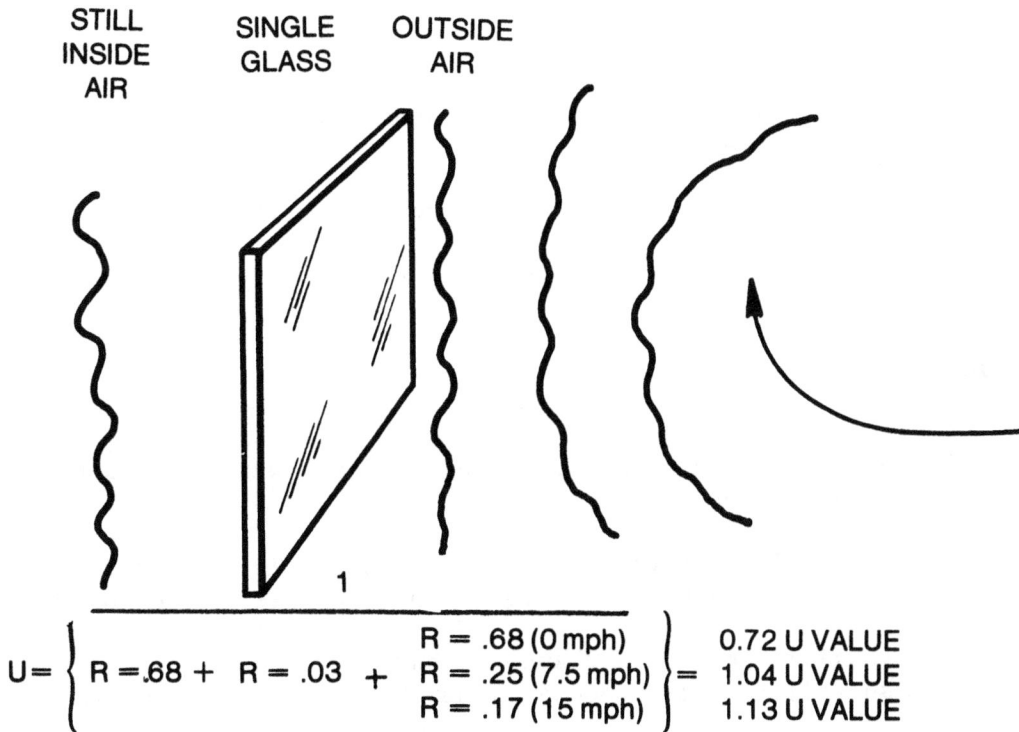

STILL   SINGLE   OUTSIDE
INSIDE   GLASS   AIR
AIR

$$U = \left\{ R = .68 + R = .03 + \begin{array}{l} R = .68 \,(0\,\text{mph}) \\ R = .25 \,(7.5\,\text{mph}) \\ R = .17 \,(15\,\text{mph}) \end{array} \right\} = \begin{array}{l} 0.72\ \text{U VALUE} \\ 1.04\ \text{U VALUE} \\ 1.13\ \text{U VALUE} \end{array}$$

Figure 3.   Heat Flow Through Glass vs. Wind Velocity

4) Prevailing winter winds come from a different direction than
    prevailing summer winds in much of the U. S. Therefore, a
    windbreak can be placed to divert winter wind away from a
    building without interfering with summer breezes.

ADVANTAGES:

1) Reduced air infiltration through cracks around windows.

2) Reduced heat loss through the glass by diminished wind erosion of the insulating boundary layer of air at the glass surface.

3) Partial protection from the summer sun on east and west orientation when the sun is low in the sky.

4) Privacy.

5) Improved natural ventilation when the windbreak geometry funnels breezes.

6) Snowdrift control (snow will collect at the leeward side).

7) Slight reduction in noise from sources beyond the windbreak.

DISADVANTAGES:

1) Possible need of pruning, fertilizing, watering, and insecticides.

2) Difficult to establish where windbreak is most likely to be effective in built-up areas due to complex wind patterns.

3) Possible increased chance of burglary when windbreak impairs surveillance of windows by neighbors or pedestrians.

AESTHETICS:

1) Windbreaks and shrubs can improve the overall aesthetic character of a neighborhood.

2) Windbreaks can be used to physically or implicitly define the boundaries between public and private space.

3) Windbreaks can obstruct distant views.

COSTS:

A three-foot-tall American Arborvitae costs $6 delivered, according to a Washington, D. C., nursery. Spaced 3 feet apart, this amounts to $2 a linear foot.

A six-foot-high fence consisting of boards staggered on either side of a 2 x 4 costs $33 per eight-foot section (including one post) delivered, according to one Washington, D. C., lumber yard. This amounts to $4.13 per linear foot.

EXAMPLES:

1) George Mattingly and Eugene Peters of Princeton University are studying the effects of wind on a group of townhouses at Twin Rivers, New Jersey. Results from scale models in a wind tunnel suggest that a five-foot-high wooden fence would reduce air infiltration 26 to 30 percent; a single row of evergreen trees as tall as the house would reduce air infiltration 40 percent; and a combination of the two would reduce air infiltration 60 percent. The best location for a windbreak was at a distance of 1 1/2 to 2 1/2 windbreak heights upwind from the house. The results of the wind tunnel tests are now being studied in a full-scale field experiment at Twin Rivers. (75,Mattingly,p37) = (year, author, page)

2) Another experiment was conducted by the Lake State Forest Experimental Station in Nebraska on two identical houses. One was fully exposed to the wind, and the other was protected by dense shrubbery. The exact fuel consumption for maintaining an indoor temperature of 70° F in each house was measured. A savings of 23 percent was recorded for the protected house. (63,Olgyay,p99)

Similar results are reported in a study conducted in South Dakota. A fully exposed electrically heated house required 443 kWh to maintain an inside temperature of 70°F from January 17 to February 17. An identical house sheltered by a windbreak required only 270 kWh. The difference in average energy requirements for the whole winter was 33.92 percent. (74,Flemer,p2)

3) Following the References are two wind maps of the U. S. showing the direction and mean velocity of wind for July and January. (68,ESSA,p73)

REFERENCES:

ASHRAE, Handbook of Fundamentals, American Society of Heating, Refrigerating and Air Conditioning Engineers, Inc., New York, 1972.

Caborn, J. M., Shelterbelts and Windbreaks, Faber and Faber, Ltd., London, 1965.

ESSA, Climate Atlas of the United States, U. S. Dept. of Commerce, Environmental Science Service Admin., Washington, D. C., June 1968.

Flemer, William III, Energy Conservation with Nature's Growing Gifts, American Association of Nurserymen, Washington, DC, 1974.

Geiger, Rudolf, The Climate Near the Ground, Harvard University Press, Cambridge, MA, 1966.

Jensen, Martin, Shelter Effect, The Danish Technical Press, Copenhagen, 1961.

Mattingly, George E. and Peters, Eugene F., Wind and Trees - Air Infiltration Effects on Energy in Housing, The Center for Environmental Studies, Princeton, NJ, 1975.

Olgyay, Victor., Design with Climate: Bioclimatic Approach to Architectural Regionalism, Princeton University Press, Princeton, NJ, 1963.

Robinette, Gary O., Plants/People/and Environmental Quality, U. S. Department of the Interior, National Park Service in Association with American Society of Landscape Architects Foundation, Government Printing Office, Washington, DC, 1972.

Rosenbrug, Norman J., Microclimate: The Biological Environment, John Wiley and Sons, New York, 1974.

Vandervort, Donald W., How to Build Fences and Gates, Lane Magazine and Book Co., Menlo Park, CA, 1971.

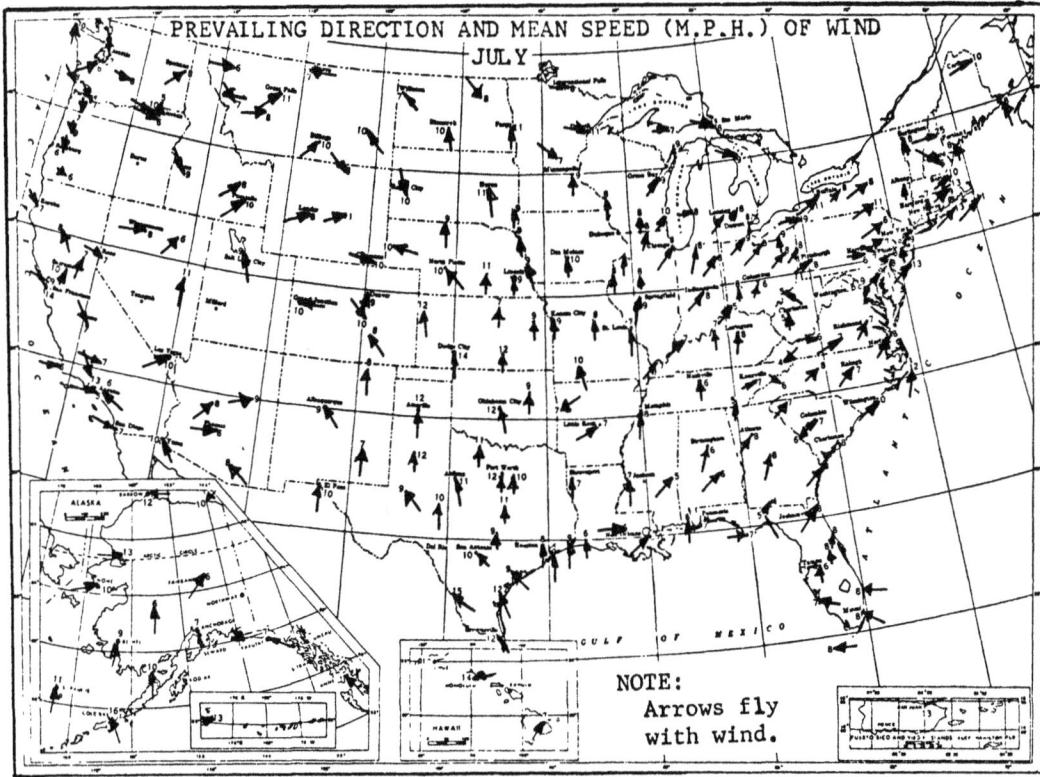

PREVAILING DIRECTION AND MEAN SPEED (M.P.H.) OF WIND
JULY

NOTE:
Arrows fly
with wind.

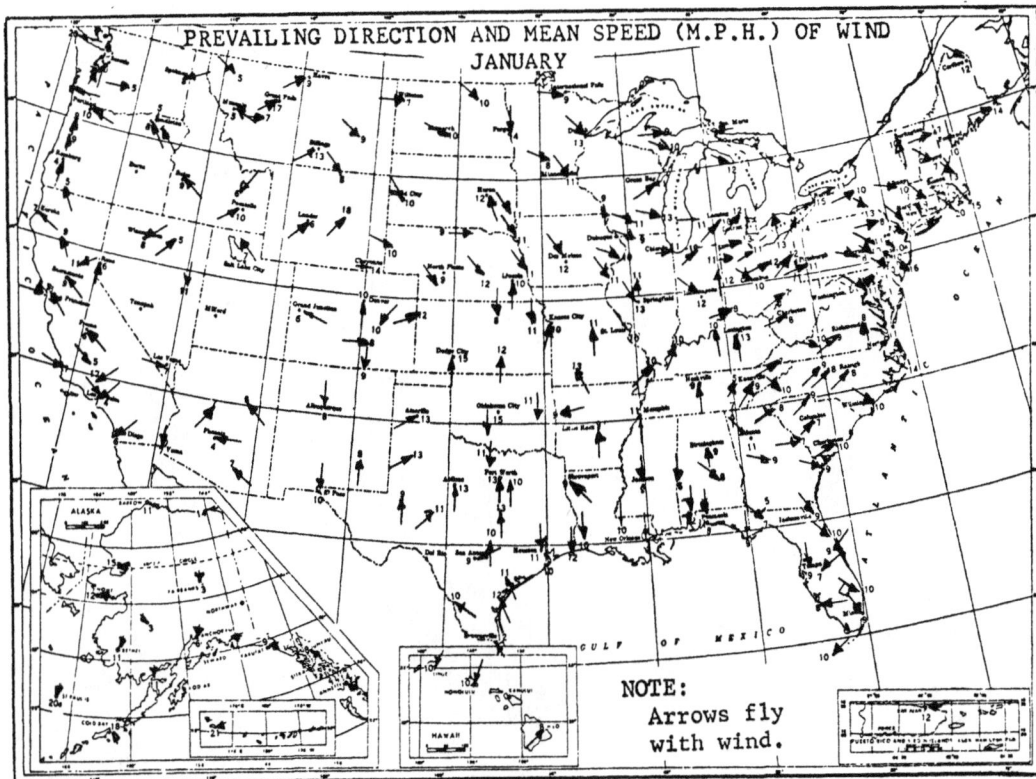

PREVAILING DIRECTION AND MEAN SPEED (M.P.H.) OF WIND
JANUARY

NOTE:
Arrows fly
with wind.

U.S. Dept. of Commerce, Environmental Science Services Admin., Washington, D.C

1.2 SHADE TREES/Shading, Insulation

STRATEGY:

> Plant deciduous trees to provide shade in the summer and admit
> sunlight in the winter. Plant evergreens to provide shade in
> the summer and to reduce window heat loss to the night sky in
> the winter.

PHENOMENA:

1) Deciduous trees provide shade in summer, then lose their
    leaves and admit sunlight in the winter. A tree-shaded,
    south-facing window receives less solar heat than an unshaded,
    north-facing window. (The north window receives diffused
    radiation from clouds.) This solar protection increases
    continuously throughout the summer, as shown below.
    (65,Forest Service,p77)

TYPE OF FOREST	PERCENT OF LIGHT PENETRATION		
	April	May	Sept.
Evergreen	8	7	4
Deciduous	51	23	5

2) Trees not only reduce window heat gain by blocking direct
    sunlight penetration but also by lowering the ground surface
    temperature. In a test conducted at Indiana University, when
    the air temperature was 84° F, concrete exposed to the sun was
    108° F while concrete shaded by a maple tree was only 88° F.
    (75,FEA,p3) The heat gain through a tree-shaded window is
    therefore diminished both by reduced heat radiation from the
    ground and correspondingly cooler air temperatures.

3) The sun's path is lower in the sky in the winter than in the summer. Therefore the sun's rays may be low enough to angle below the branch structure of a tree adjacent to a window.

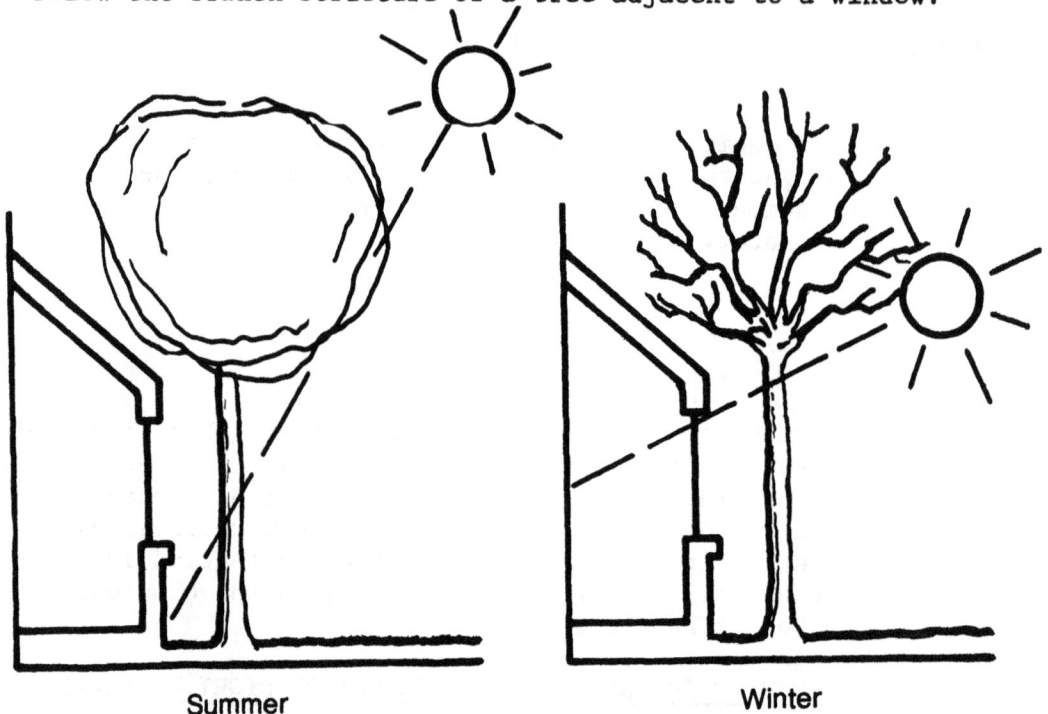

Summer   Winter

Figure 4.   Seasonal Sun Angle vs. Tree Interference

4) On a winter night the outside surface of the window radiates heat to the terrain and the sky. The winter sky has a much colder average temperature compared to the ground which re-radiates heat accumulated during the hours of sunlight.

An evergreen tree in close proximity to a window will obstruct the winter night sky. The temperature of the tree will approximate the air temperature. Radiant heat loss from the window to the tree will therefore be less than to the much colder night sky. This phenomena is illustrated by frost appearing on open fields earlier than under trees.

ADVANTAGES:

1) Reduced summer solar heat gain with only slightly decreased winter solar heat gain when deciduous trees are used. (Trees with foliage low to the ground are especially effective on east and west exposures where roof overhangs provide no protection from low angle sunrays.)

2) Reduced summer solar heat gain and reduced winter night-sky heat losses with the use of evergreens.

3) Reduced or eliminated glare from the sun and bright sky.

4) Protection of windows from driving rain or hail.

DISADVANTAGES:

1) Required maintenance, including not only fertilizing, pruning deadwood, and possibly spraying to control insects, but also removal of leaves from grounds and rainwater gutters.

2) Increased likelihood of storm damage due either to lightning or wind-broken tree limbs.

3) Possibility of root blockage of underground sewer pipes.

4) Beneficial winter solar heat gain blocked by evergreens.

AESTHETICS:

1) The quality of light under trees is much different from the quality of light under a roof overhang. Light under a roof overhang is principally "blue light" diffused from the blue sky. The light under deciduous trees is principally "red light" filtered through the leaves. This effect is greater with deciduous trees than coniferous trees. (65,Forest Service,p65). In addition to the difference in color quality, trees may dapple a window with a pattern of sunflecks penetrating the canopy of leaves or needles. This pattern moves with breezes and changes density with the seasons.

2) Trees may be selected for their softness and irregularity or their natural or pruned shapes to complement the hard-edged geometry of the building. For example, maple and ash provide a circular massing in summer and an ascending open branch pattern in winter. The linden is spherical also, but in winter it has a dense, twiggy branch pattern. Honey locust and tulip trees are vertical oblongs in form, while white oaks are horizontal oblongs. Poplars are column-like in shape and American elms are vase-shaped. (73,Olgyay,p76)

3) Trees can provide an effective unifying element to a complex of buildings.

4) Trees can affect the scale of the setting in which a building is seen.

COST:

A Washington, DC, nursery quoted the following prices for
trees delivered in small quantity (price does not include
planting). A 6-foot oak or maple costs $8 to $10. Prices of
taller trees vary with their growth rate: a 20-foot maple
costs $70, a 20-foot oak costs $300 to $350. A 6-foot hemlock
costs $25, a 10-hemlock costs $50.

EXAMPLES:

The following diagram illustrates the calculated effectiveness
of a shade tree on the east or west side of a house in reducing
the air temperature in the shaded area.   (74,Weatherwise
Gardening,p32)

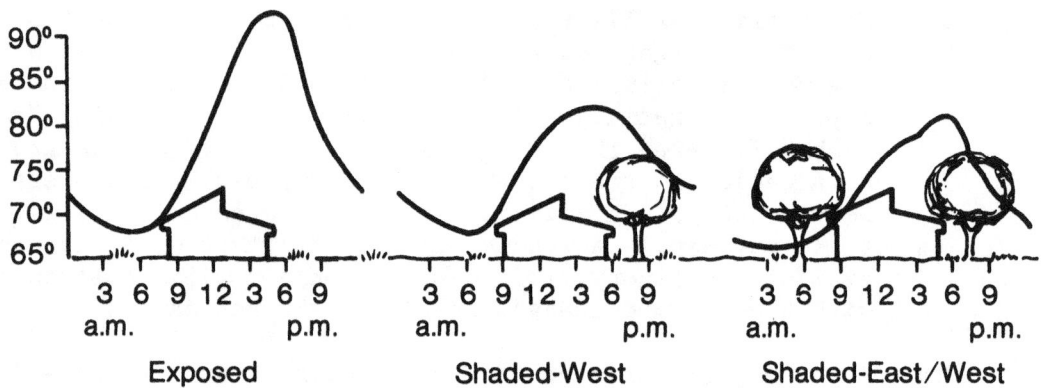

Figure 5.   Shade Tree Effectiveness vs. Orientation

REFERENCES:

American Society of Landscape Architects, Site Planning for Solar Energy Utilization.

FEA, Smith Family Used Living Insulation, Energy Reporter, Federal Energy Administration, Washington, DC, Aug/Sept. 1975

FEA/NBS, Energy Conservation with Landscaping, Federal Energy Administration, Washington, DC, 1976.

Forest Service, Radiant Energy in Relation to Forests, Tech. Bul. 1344, U. S. Department of Agriculture, Washington, DC, Dec. 1965.

Olgyay, Victor, Design with Climate, Princeton Univ. Press, NJ, 1973.

Reifsnyder, William E. and Lull, Howard W., Radiant Energy in Relation to Forests, Tech. Bul. No. 1344, Forest Service, Washington, DC, Dec. 1965.

Weather-Wise Gardening, Ortho Book Division, Chevron Chemical Company, San Francisco, CA, 1974.

1.3   GROUND SURFACES/Daylighting, Solar Heating

STRATEGY:

      Use light-colored ground surfaces to reflect sunlight into
      windows, dark-colored surfaces to absorb sunlight and raise
      outside temperatures, or planted surfaces to absorb both
      sunlight and lower outside temperatures.

PHENOMENA:

1)   Light reflected from the ground represents 10 to 15 percent
    of the total daylight transmitted by a first floor window on
    the sunlit side of a building, and may account for more than
    half of the total daylight on the non-sunlit side. The amount
    of light reflected through the window is even greater when
    adjacent ground surfaces are light in color. The following is
    a list of common ground surfaces and the percentage of inci-
    dent light they reflect. (72,IES,p75)

MATERIAL	PERCENT REFLECTED
White paint (new)	75 percent
(old)	55
Snow (new)	74
(old)	64
Concrete	55
Marble (white)	45
Granite	40
Brick (buff)	48
(dark glazed)	30
Vegetation (average)	25
Macadam	18

2) Ground-reflected light transmitted through windows strikes the ceiling. This is beneficial for daylighting in two respects. First, the light is projected deeper into the room than is direct sunlight. Second, ceilings are usually light-colored and, hence, reflect light better than darker floors, carpets, or furniture.

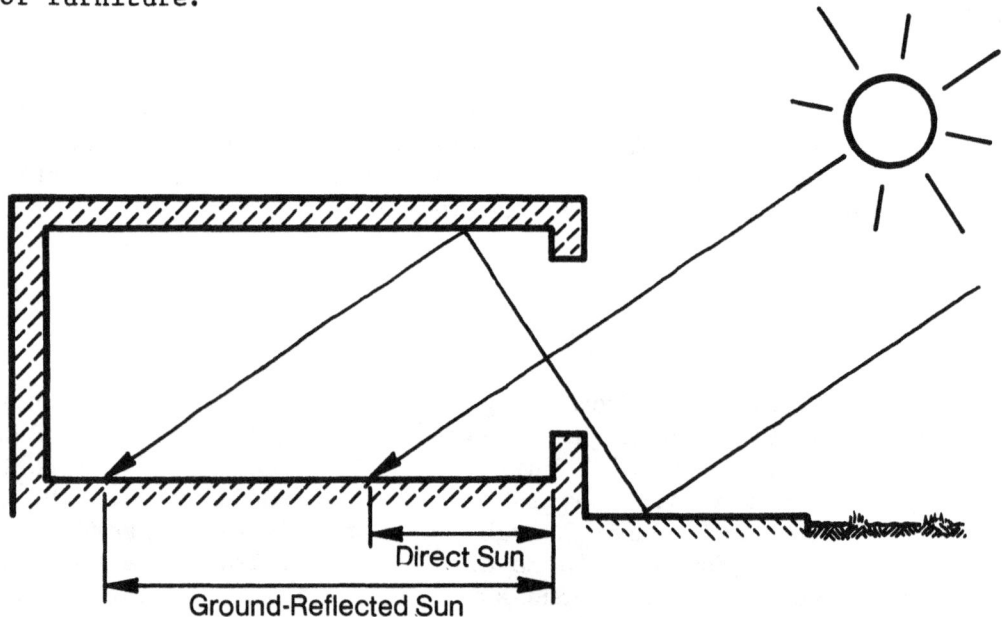

Direct Sun

Ground-Reflected Sun

Figure 6. Ceiling Reflected Ground Light

The addition of ground reflected light to direct sunlight increases the ability of a window to provide supplemental winter heat. (The additional light becomes additional heat.)

3) Dark-colored surfaces absorb more light than light-colored surfaces and therefore become warmer in sunlight. A window will radiate less heat in the winter when adjacent ground surfaces are warm. Also, on a calm winter day, the air temperature over dark-colored ground surfaces will be warm, further reducing window heat losses. Conversely, during the summer, light-colored surfaces are beneficial because they absorb less light than dark-colored surfaces and are con- sequently cooler. The following is a list of common ground surfaces and their sunlit surface temperatures when the air temperature is 84° F. (75,FEA,p4)

MATERIAL	SURFACE TEMP.	DEV. FROM AIR
Dark Asphalt	124	+40
Light Asphalt (dirty)	112	+28
Concrete	108	+24
Short grass (1-2 inches)	104	+20
Bare ground	100	+16
Tall grass (36 inches)	96	+12

4) Plant cover absorbs sunlight, yet has a lower surface temperature than paving. Evaporative cooling occurs during the transpiration life process of plants. The net heat gain from the sun is rapidly dissipated by the enormous surface area of leaves. Very little heat is stored in vegetation because of its minimal mass. Night air temperatures over grass, for example, are therefore cooler than over paving. The lower day temperatures and lower night temperatures of planted surfaces result in less window heat gain and a reduced air conditioning burden compared to the situation of having paved surfaces adjacent to windows.

ADVANTAGES:

1) Increased daylight penetration when light-colored ground surfaces occur adjacent to windows.

2) Reduced window heat loss in the winter when dark-colored ground surfaces occur adjacent to windows.

3) Reduced heat gain in the summer, both day and night when planted surfaces occur adjacent to windows. Likelihood of glare reduced also.

DISADVANTAGES:

1) Increased light admitted into the building. When absorbed, the resulting heat is a disadvantage in summer.

2) Increased likelihood of glare with light-colored ground surfaces.

3) Ineffectiveness of ground surfaces to improve the performance of windows several stories above the ground.

4) Increased reflection of sound through windows when hard paved surfaces are used.

AESTHETICS:

1) Brick, cobblestone, asphalt paving blocks, gravel, stone slabs, and textured concrete are a few of many options which can be used in a multitude of patterns to reinforce the geometry of a building scheme.

2)   Lawn, ground cover, or shrubbery strategically placed to
     reduce the ground heat and glare adjacent to windows can
     enhance both the view out and provide an attractive setting
     for a building.

EXAMPLES:

1)   T. Kusuda, at the National Bureau of Standards, measured the
     surface temperature of five different ground surfaces (bare
     soil, black top, long grass, short grass, and white paint
     over black top) for two years. He found that asphalt had an
     average yearly temperature 8° F higher than grass. The daily
     maximum surface temperatures of the asphalt paving reached
     140° F, whereas the bare soil seldom exceeded 100° F. Even
     during the morning on an average summer day, asphalt surfaces
     were warmer than ambient air temperatures, while bare soil was
     10° F cooler than ambient air temperatures. Painting the
     asphalt paving white reduced its surface temperature con-
     siderably, even during the second year when the paint had
     faded. Temperatures of the painted surface never exceeded
     105° F. The following graph gives a breakdown of surface
     temperatures of different surfaces by month.   (76,Kusuda,p297)

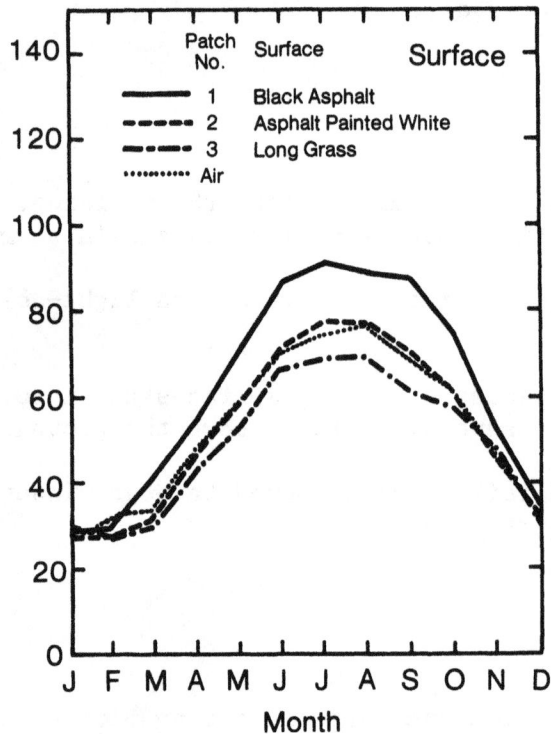

Figure 7. Monthly Ground Surface
Temperatures

2) Measurements have been made comparing temperatures over Merion Blue Grass, artificial turf and asphalt paving. When the air temperature was 90° F, relative humidity 40 percent, and wind 1.1 mph, the grass was 100° F, the asphalt paving 140° F, and the artificial turf was 162° F. It was expected that grass would be cooler than artificial turf because of evaporative cooling. While this was the case, the temperature difference was also influenced by the light absorption characteristics of the different surfaces. The absorption of sunlight was: by grass, 78.4 percent; by asphalt, 87 percent; and by artificial turf, 92.7 percent. The grass is therefore cooler, not only due to evaporative cooling but also due to the fact that it reflects almost three times more sunlight than artificial turf and nearly two times more sunlight than asphalt paving. (71,Taylor,p2-43) The following figure shows how much warmer the air was at various heights above artificial turf compared to Merion Blue Grass.

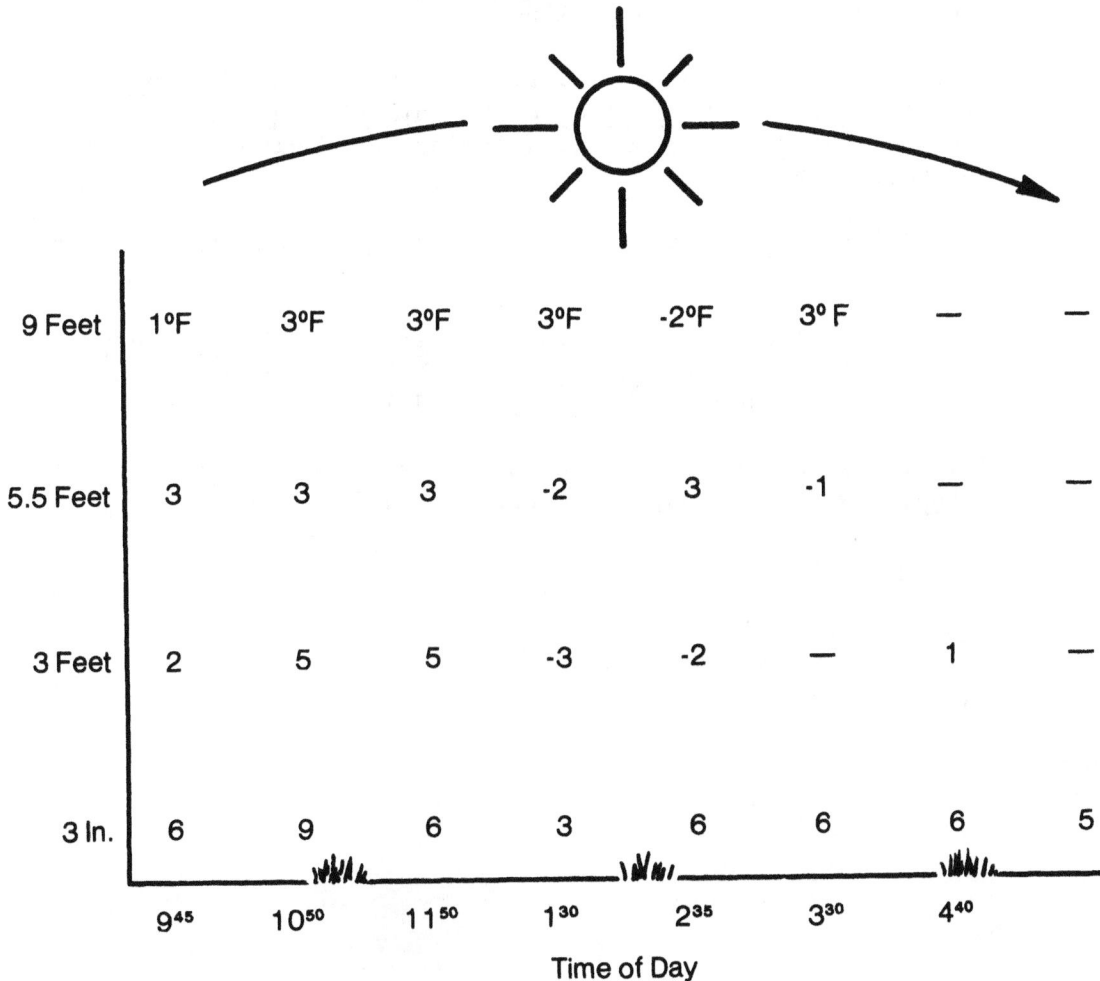

	9⁴⁵	10⁵⁰	11⁵⁰	1³⁰	2³⁵	3³⁰	4⁴⁰	
9 Feet	1°F	3°F	3°F	3°F	-2°F	3° F	—	—
5.5 Feet	3	3	3	-2	3	-1	—	—
3 Feet	2	5	5	-3	-2	—	1	—
3 In.	6	9	6	3	6	6	6	5

Time of Day

Figure 8. Air Temperature Differences Between Artificial Turf and Merion Blue Grass

3) A terrace surfaced in a dark material, facing south and east on the inside corner of an L-shaped building, is a very effective sun pocket. A temperature of 70° F (21.1° C) has been recorded in such a corner, while out in the wind the temperature read 30° F (-1.1° C). (50, Fitch,p97).

REFERENCES:

FEA, Smith Family Uses Living Insulation, Energy Reporter, Federal Energy Administration, Washington, DC, Aug/Sept. 1975.

Fitch, James Marsten, American Building, Houghton Mifflin Publ., Boston, 1966.

Geiger, Rudolf, The Climate Near the Ground, Harvard University Press., Cambridge, 1966.

Griffeth, J. W., Wengler, O. F. and Conover, E. W., The Importance of Ground Reflection in Daylighting, Illuminating Engineering, Vol. 48, No. 1, Illuminating Engineering Society, New York, Jan. 1953.

IES, IES Lighting Handbook, Illuminating Engineering Society, NY, 1972.

Kusuda, T., The Effect of Ground Cover on Earth Temperature, The Use of Earth Covered Buildings, NSF-RA-760006 National Science Foundation, Washington, DC, July 1975.

Taylor, Elwynn and Pingel, Gerald, Green Grass That's Not So Green, New York Times, NY, April 11, 1971.

1.4 ORIENTATION TO SUN/Solar Heating, Shading

STRATEGY:

> Provide the largest window area on the side where the sun
> exposure minimizes combined mechanical heating and cooling
> needs.

PHENOMENA:

1) Sunlight transmission through windows will be a net benefit on
   an annual basis if winter solar heat gain exceeds winter
   window heat loss and summer solar heat gain.

2) What percentage of the incident solar energy a window transmits for any given day depends upon the angle at which rays of
   sunlight intercept the window and how many hours the window
   receives sunlight. The angle at which sunlight intercepts the
   window affects the amount of solar energy transmitted in two
   ways: by determining the proportion of light reflected,
   absorbed and transmitted; and by determining the projected
   area of the window measured perpendicular to the rays of
   light.

   The proportion of light reflected or absorbed increases
   gradually from the minimum when the angle between the light
   rays and the glass is 90° up to approximately 45°. Thereafter, the amount reflected or absorbed increases drastically
   until no light is transmitted. (See Strategy: Window Tilt)

   The projected area is the area of a window projected onto a
   plane perpendicular to the rays of light. The projected area
   becomes smaller as the angle at which the light intercepts the
   glass becomes smaller. The amount of light transmitted is
   therefore reduced due to the decreased area exposed as shown
   below.

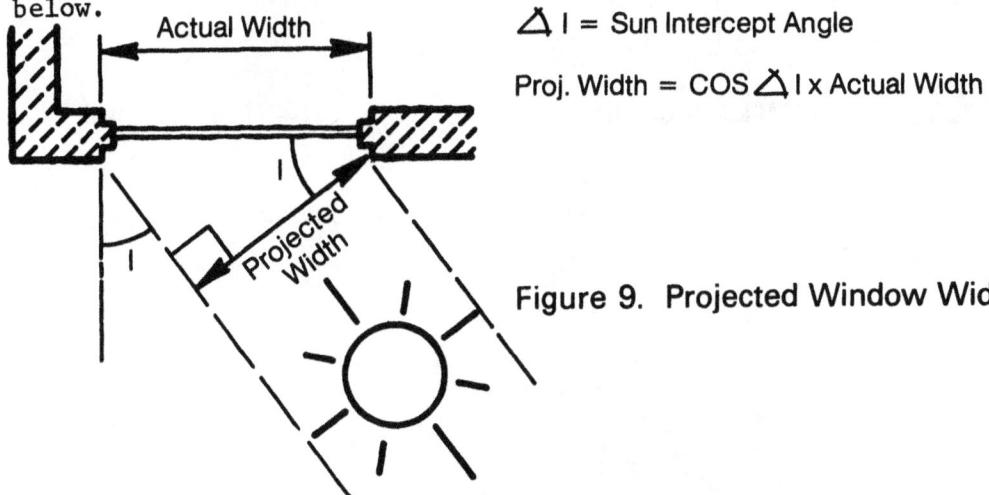

$\triangle$ I = Sun Intercept Angle

Proj. Width = COS $\triangle$ I x Actual Width

Figure 9. Projected Window Width

3) How many hours a window receives direct sunlight, as well as the angle of the rays of sunlight, is determined by the path of the sun across the sky. The path of the sun varies with the seasons and with the latitude. The summer sun rises north of east and sets north of west. The winter sun rises south of east later in the morning, travels in a lower arc, and sets south of west earlier in the evening. The further north a site is, the greater are the seasonal northerly and southerly shifts of sunrises and sunsets, the lower the arc of the sun across the sky, and the more hours the sun is above the horizon in summer and the fewer hours it is above the horizon in winter. This phenomenon is illustrated below for latitudes of 42° and 34°. (50,AIA,p35)(51,AIA,p2-16)

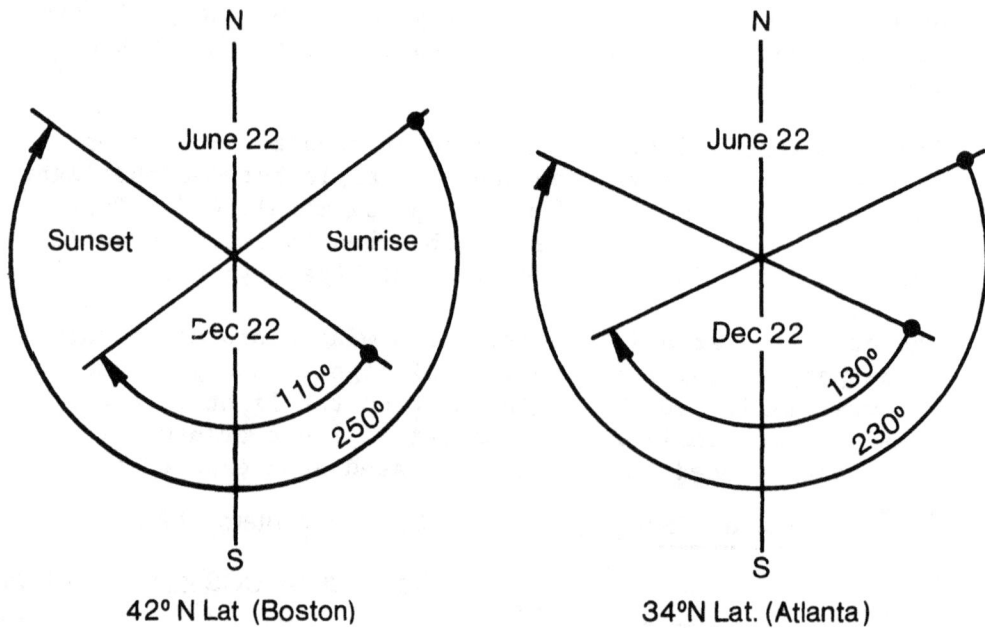

Figure 10. Plan View of Sunrise and Sunset

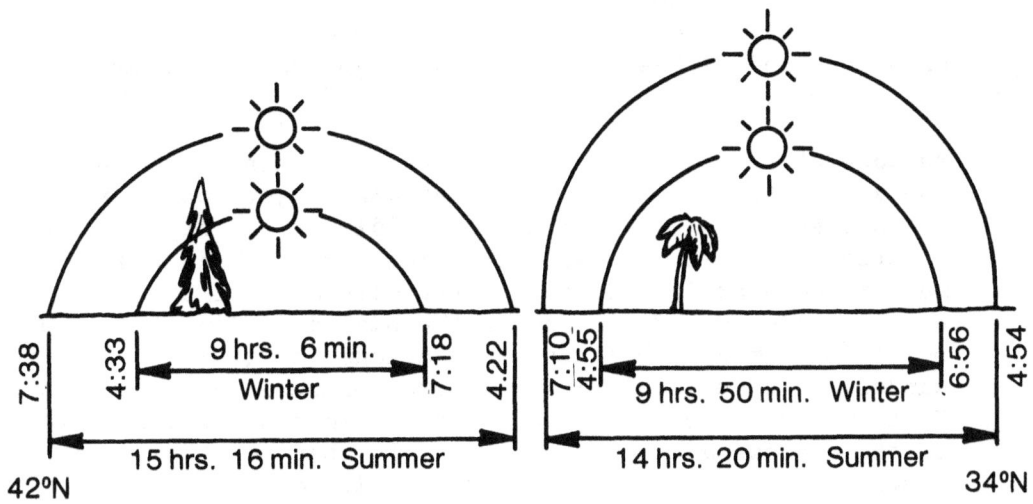

Figure 11. Daily Path of the Sun Viewed Looking South

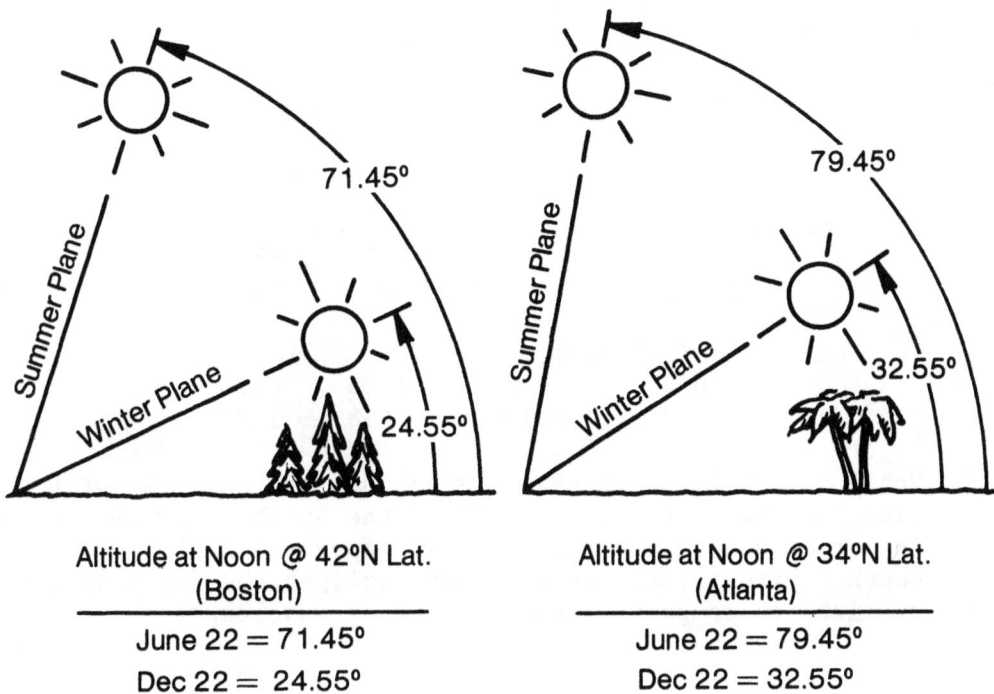

Altitude at Noon @ 42°N Lat. (Boston)	Altitude at Noon @ 34°N Lat. (Atlanta)
June 22 = 71.45°	June 22 = 79.45°
Dec 22 = 24.55°	Dec 22 = 32.55°

Figure 12. Daily Path of the Sun Viewed Looking East

4) Because of the lower and more southerly path of the winter sun across the sky, a south-facing window receives sunlight at a more direct angle and for more hours of the day than an east- or west-facing window. A south-facing window receives the sun obliquely at sunrise, almost perpendicularly at noon and again obliquely at sunset. By comparison, an east- or west-facing window receives the sun obliquely and for less than half the number of hours the sun is above the horizon. A north-facing window receives no direct sun in the winter and, hence, has only the minimal heat gain from daylight to offset conducted heat losses.

Because of the higher and more northerly path of the summer sun across the sky, a south-facing window gets no direct sun at sunrise or sunset, and in the middle of the day the sun intercepts the windows at a glancing angle. This results in a much reduced projected window area with much of the light being reflected. Furthermore, the high position of the summer sun permits modest building projections to totally shade the window. (While permitting the lower winter sun to penetrate.) East and west exposures receive the summer sun for more hours of the day and at a more direct angle than south exposures. Hence, east and west are more difficult to shade. The north exposure also receives summer sun, but for only a short period of the day and at very oblique angles. The thermal consequences of north-facing windows in the summer are therefore minimal. The solar gains for the different orientations are shown below. (76,Kusuda)

AVERAGE DAILY BTU/SQ FT/DAY

	S	E or W	N
Lat 42° N (Boston)			
Jun 22	786	1026	638
Dec 22	757	286	143
Lat 34° N (Atlanta)			
Jun 22	681	1105	681
Dec 22	1050	458	220

From these values, it can be seen that the further north a site is, the more winter sunlight the south exposure receives in comparison to the east or west exposures. The further south a site is, the less summer sunlight the south exposure receives in comparison to the east or west exposures.

Therefore, buildings located in southern latitudes should have window areas concentrated on the north and south exposures (ideally with a projecting, horizontal shading device over south-facing windows) to minimize the air conditioning burden. To obtain the greatest benefit from the sun as a winter heat source, buildings located in northern latitudes should have window areas concentrated on the south (with minimal window areas to the north).

5) The arrangement of rooms relative to their window orientation will determine their natural daily temperature cycle. By matching the times rooms are likely to be occupied with the hours they receive sunlight, solar heating can be better utilized and dependence on mechanical heating reduced.

ADVANTAGES:

1) Increased winter solar heat gain. Properly sized and oriented windows can gain more heat from the sun than they lose by conduction. (75, Berman) This can translate into reduced heating costs.

2) Decreased summer solar heat gain. Properly oriented and shaded windows can result in a savings in initial cost and subsequent operation of one ton of air conditioning per 100 square feet of glass compared to poorly oriented unshaded windows. ('66,Callender,p.749)

DISADVANTAGES:

1) Fading fabrics due to exposure to sunlight.

2) Winter overheating possible with large south-facing windows for small rooms within light-weight construction buildings.

3) Reflected sunlight from light-colored ground surfaces or adjacent buildings may reduce effectiveness of glass orientation relative to direct sunlight.

AESTHETICS:

Properly oriented windows will result in rooms having a shaded character in summer and a sunny, bright character in winter.

The orientation of a window will determine the path of the patch of direct sunlight projected onto the floor and/or walls. The patch of sunlight may be attractive if it spotlights a geranium, or bothersome if it results in glare on a work surface.

The orientation of a window will determine whether the outward view is shaded or sunlit. A window oriented to admit the sun (e.g., south-facing) provides a view of the shaded side of outside objects. A window on the shaded side of a building (e.g., north-facing) provides a view of the sunlit side of outside objects. East or west-facing windows offer the advantage of a view with the light source changing direction between morning and afternoon.

COSTS:

Proper orientation of windows does not presume increasing the total window area. The issue is the distribution of the window area. Therefore, proper orientation need not result in additional construction costs.

Proper orientation of windows may even reduce construction costs, by minimizing the amount of expensive shading required by adverse orientation, and reducing the required capacity of the heating and cooling mechanical systems.

EXAMPLES:

1) The following is a study of the effect of window orientation on heating costs. The window areas for a conventional house in Boston are revised as follows:

WINDOW AREA IN FT^2

	S	N	E	W	TOTAL
CONVENTION	100	100	50	50	300
REVISED	180	20	50	50	300

Occupancy was considered to be a family of four, lighting and appliance heat gains were considered at about 20 kWh per day, and room temperature was to be maintained at 70° F (21° C). Solar data were taken from ASHRAE and outside temperature data from the U. S. Weather Bureau. U-values for the building envelope were FHA minimum: Roof 0.053, walls 0.085, doors 0.65, floors 0.084, windows 0.65. Air infiltration was assumed at 1 change per hour.

The conventional house was calculated to require 92 million Btu per year from the heating system. By merely shifting 80 square feet of windows from the north to the south sides of the house, the heating requirement was reduced to 83 million Btu for a net savings of 9 million Btu. (76,Bliss,p34)

2)   The cooling loads resulting from different window orientations have been studied with scale models at the Building Research Station at Haifa, Israel. In one experiment, four identical models were constructed. The walls consisted of light-weight concrete 150 mm (5.9 in) thick. One side contained a closed window. The models were oriented so that the windows faced each of the four cardinal directions. Before sunrise, the inside air temperatures were all approximately equivalent.

The model with the west-facing window had the largest inside-to-outside temperature difference. The difference reached 11° C (19.8° F) in the afternoon.

Four hours after sunrise, the inside air temperature of the east-facing model had risen 8° C (14.4° F) above the outside air temperature.

The model with the west-facing window had the largest inside-to-outside temperature difference. The difference reached 11° C (19.8° F) in the afternoon.

The models with south- and north-facing windows showed the smallest inside/outside temperature differences ranging from 3° to 5° C (5.4° to 9° F). (69,Givoni,p201)

3)   Measured data documenting the effect of orientation on inside temperatures of actual buildings were collected in a study by the city of Davis, California, in 1974. The energy require-ments of a newly constructed but not yet occupied apartment complex were recorded. The apartments were two- and three-story buildings built to conventional standards and oriented north, south, east, and west.

The study found that top floor apartments facing east or west were the hottest, reaching a maximum of 99° F (37.2° C). The same apartment model facing north or south reached a maximum temperature of 85° F (29.4° C). (76,Cole,no pages)

4)   The following graph by Victor Olgyay differentiates total
     solar gain during the cooling season (overheated period)
     from total solar gain during the heating season (underheated
     period) in New York City. (73,Olgyay,p58)

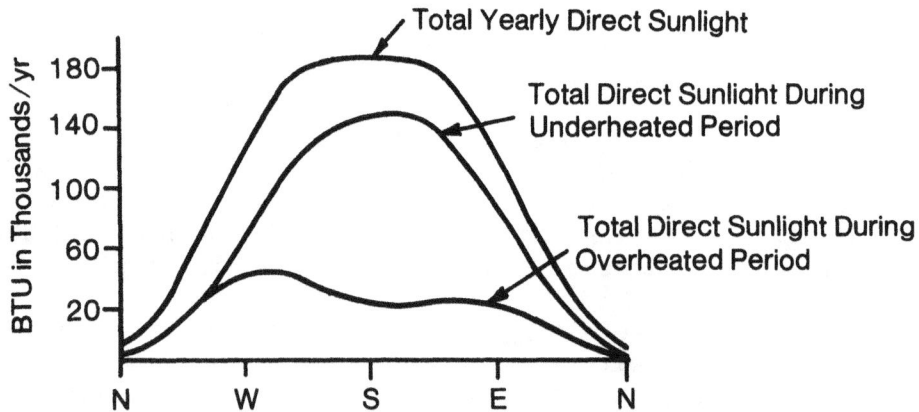

Figure 13.   Orientation vs. Yearly Solar Radiation in New York

The most favorable orientation occurs where the positive
difference (the solar gain during the cooling season subtracted
from solar gain during the heating season) is greatest. For
New York City this is 17 1/2 degrees east of south.

5)   A study by Samuel M. Berman of Stanford University provides
     data quantifying the yearly effect of solar energy on the
     energy balance of a window. Calculations were made for several
     cities and considered a variety of window glazing and shading
     options. The example shown below assumes the window includes
     a storm sash or is double glazed, and that, during the summer,
     a standard white window shade or venetian blind is lowered
     when the window is in sunlight. No external shading of the
     window is assumed. The calculations show the amount of
     indoor climate control energy expenditure due to the window in
     KBtu per square foot of window for the season. A positive
     value represents a net energy gain, a minus value represents
     a net energy expenditure. (75,Berman,p65)

	SOUTH		EAST & WEST		NORTH	
	Winter	Summer	Winter	Summer	Winter	Summer
DALLAS-FT. WORTH	+107	-41	+48	-61	+9	-34
NEW YORK CITY	+71	-18	+14	-24	-25	-13

The above values show that the solar benefit in the winter more than offsets the solar detriment in the summer for south-facing windows in regions with hot summers such as Dallas, or cold winters like New York. Also noteworthy is the fact that east and west-facing exposures nearly break even. Finally, it must be realized that the net gain or loss of a window may be advantageous or detrimental, depending on the configuration of the mechanical heating and cooling systems.

REFERENCES:

AIA, <u>Regional Climate Analysis and Design Data</u>, The House Beautiful Climate Control Project, Bulletin of the AIA, Washington, DC, 1950 & 1951.

Berman, Samuel and Silverstein, Seth, <u>Energy Conservation and Window Systems</u>, National Technical Information Service, Springfield, VA, Jan. 1975.

Bliss, Raymond W., and Williams, Robert H. (editor), <u>Why Not Just Build the House Right in the First Place?</u>, Bulletin of the Atomic Scientists. Educational Foundation for Nuclear Science, Chicago, IL, 1976.

Callender, John Hancock, <u>Time Saver Standard</u>, McGraw Hill, NY, 1966.

Cole, Robert S., <u>Solar Cost Reduction Through Window and Building Orientation</u>, <u>Solar Engineers</u>, Solar Engineering Publishers, Inc., Dallas, TX, April 1976.

Givoni, B., <u>Man, Climate and Architecture</u>, Elsevier Publishing Company, New York, NY, 1969.

Griffith, J. W., <u>Design of Windows</u>, <u>Solar Effects on Building Design</u>, Publication No. 1007, BRI, Inc., Washington, DC, 1963, pp. 95-101.

Knowles, Ralph C., <u>Energy and Form</u>, The MIT Press, Cambridge, M, 1974.

Kusuda, T., <u>Calculations</u>, National Bureau of Standards, Washington, DC.

Markus, Thomas A., <u>Window Design in Europe: A Review of Recent Research</u>, <u>Solar Effects on Building Design</u>, Publication No. 1007 BRI, Inc., Washington, DC, 1963, pp. 119-140.

Nickerson, David. <u>Best Ways to Gain From a Solar House</u>, Washington Post, Washington, DC, Sept. 5, 1976.

Olgyay, Victor, <u>Design with Climate</u>, Princeton University Press, Princeton, NJ, 1973, p. 53.

Peter, John, <u>Design with Glass</u>, Reinhold Publishing Corp., New York, NY, 1964.

Total Environmental Action, Inc., <u>Solar Energy Home Design in Four Climates</u>, Total Environmental Action Press, Harrisville, NH, 1975.

1.5 ORIENTATION TO WIND/Ventilation, Air Tightness

STRATEGY:

Provide "cross-ventilation" with windows out of alignment
with the direction of the wind to improve overall ventilation.

PHENOMENA:

1) When window placement on opposite sides of an interior space
is possible, the building should be oriented slightly askew to
the direction of the wind. When window placement on opposite
sides of a space is not possible but placement on adjacent
sides is possible, the building should face directly into the
wind. The reason for this is illustrated in the following
plan view of window locations:

Windows on Opposite Sides

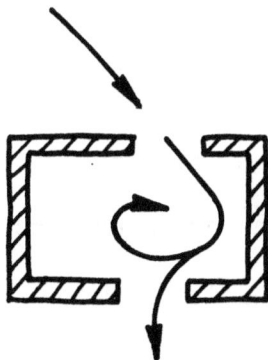

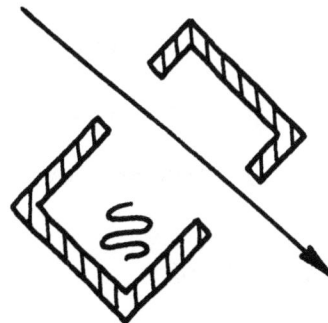

Good Overall Circulation   Local Circulation

Windows on Adjacent Sides

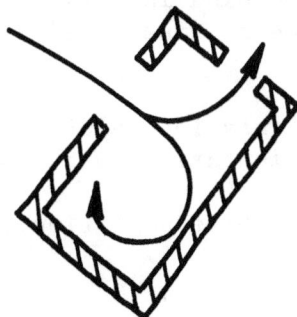

Good Overall Circulation   Local Circulation

Figure 14.   Window Location vs. Air Circulation

2) From this figure, it can be seen that if the wind encounters
    an inlet and outlet in alignment with its outside direction it
    will pass through the intervening space in a narrowly defined,
    high velocity stream. Very little ventilation will occur
    beyond that narrowly defined stream. However, if the wind is
    forced to change direction in transit between inlet and outlet
    a turbulence within the room will develop. A circular current
    will encompass the sides and corners of the room. The maximum
    air speed is reduced compared to windows in direct alignment
    with the wind, but the average velocity of air movement within
    the entire space will be greater. Overall ventilation is
    subsequently superior.

3) Where the building interior is subdivided into a series of
    interconnected spaces, placement of interior partitions can
    provide the disruption of the otherwise straight path of air
    flow between upwind and downwind windows.

ADVANTAGES:

1) Increased volume of room ventilation when windows properly
    located.

2) Decreased annoyance from local high velocity drafts.

3) Reduced demand for air conditioning and mechanical ventilation

DISADVANTAGES:

1) Effectiveness unreliable if wind direction is unpredictable
    and subject to wide variation.

2) Diminished effectiveness if the wind is very weak. Air cir-
    culation within a space will be so dissipated as to be beyond
    perception.

3) Potential conflict between orientation relative to the wind,
    orientation to the sun, and orientation to the view.

4) Untenable where extreme air or noise pollution prevails.

AESTHETICS:

1)   Opened window may provide pleasant sounds and smells to enter a room, providing variation to the quality of the interior.

2)   A strategically directed breeze through a window may be more psychologically effective than an equal rate of air changes per hour ducted through a register.

3)   Open windows generally require insect screens, which affect the exterior character of the fenestration.

COSTS:

The cost of locating windows on opposite or adjacent sides of an interior space is not likely to be higher than locating the same number of windows on only one side of the space.

EXAMPLES:

The following table illustrates the calculated effect of window location relative to the wind upon the average inside air velocities, given as a percent of the outside wind velocity.   (69, Givoni, p261)

	Wind Perpendicular	Wind at an Angle
Windows on opposite sides	35%	42%
Windows on adjacent sides	45%	37%

REFERENCES:

Givoni, B., Man, Climate and Architecture, Elsevier Publishing Co., New York, NY, 1969.

Olgyay, Victor, Design with Climate: Bioclimatic Approach to Architectural Regionalism, Princeton Unit. Press, Princeton, NJ, 1963.

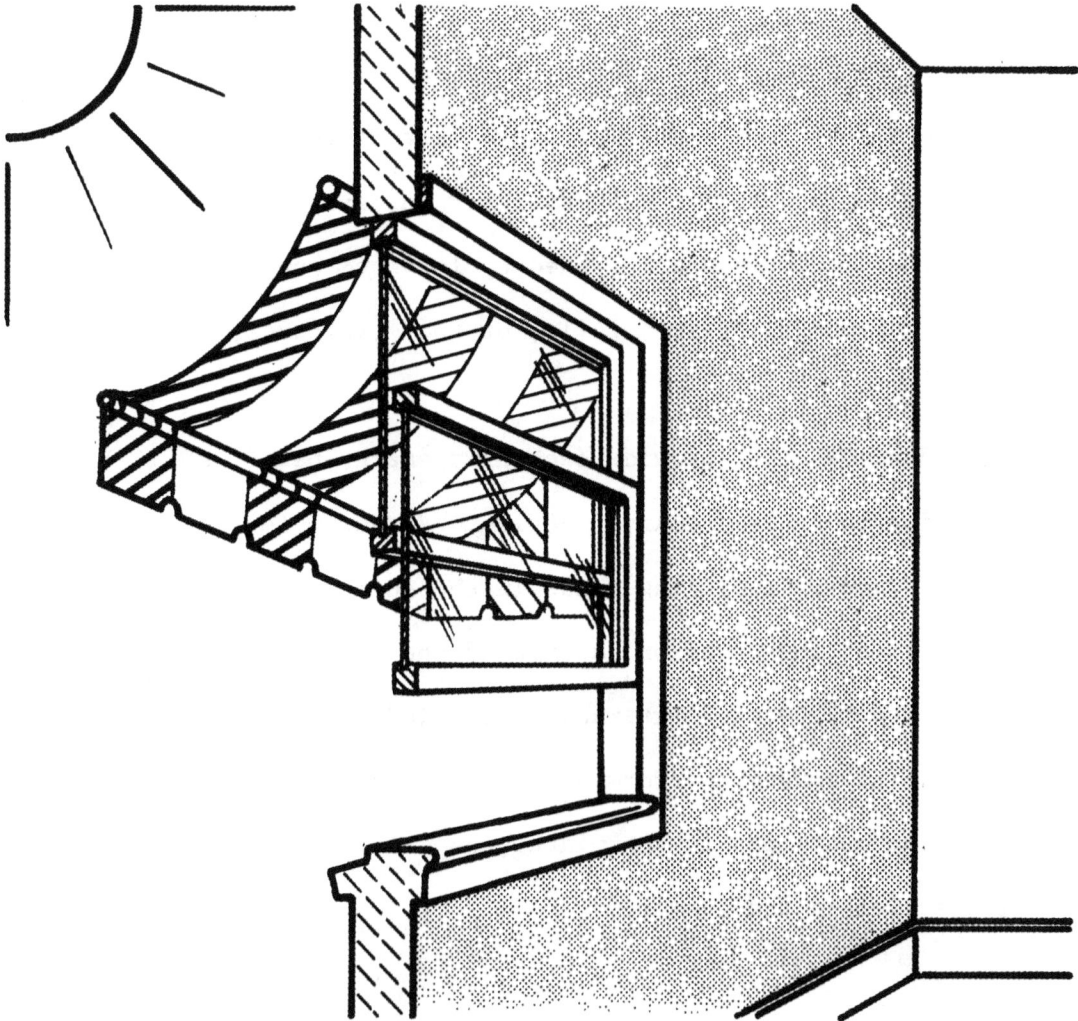

EXTERIOR APPENDAGES

SUN SCREENS
EXT. ROLL BLINDS
ARCHITECTURAL PRO-
JECTIONS
EXT. SHUTTERS
AWNINGS

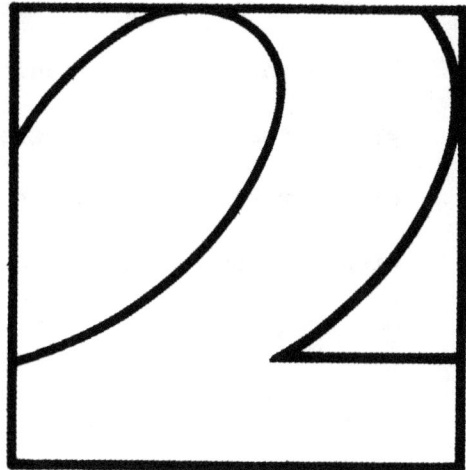

2. EXTERIOR APPENDAGES

The energy performance of a window can be greatly improved with external appendages which are part of the window system or part of the wall or roof system. For example, shading can be effectively accomplished by louvered sun screens, blinds, awnings, brise soleils, or roof overhangs.

Such devices often provide secondary energy benefits in addition to their primary function. For example, sun screens in addition to providing shade in the summer, preserve the air film at the exterior surface of the glass, thus reducing winter heat losses. External roll blinds, by providing a trapped air space, perform similarly to storm windows in reducing heat loss in the winter, as well as affording sun protection in the summer.

The general advantage of using exterior appendages to improve window performance is that they mitigate climatic problems before they enter the building. Also, although to a lesser extent than site strategies, external appendages allow some of the residual forces, such as summer solar heat or winter winds, to be dissipated before encountering the window.

STRATEGY:

Install a screen of mini-louvers outside a window to shade
direct sunlight, yet provide a view out from inside the
building.

PHENOMENA:

1) The effectiveness of a solar screen in shading a window depends
   on its geometry and its reflectivity as a material. The
   geometry determines how high the sun must be above the horizon
   before the louvers block all the direct sunlight. The reflectivity of the louvers determine how much light penetrates
   indirectly by being reflected off the surface of the louvers.
   If the slats have a reflective surface, part of the light
   striking the top of one slat will be reflected directly
   through the window and part will be directed to the underside
   of the slat above, and then directed through the window.
   Thus, highly absorptive surfaces improve the effectiveness of
   sunscreens. The following is an example of the geometry and
   effectiveness of a sun screen reported by one manufacturer.
   (76,Koolshade,p6)

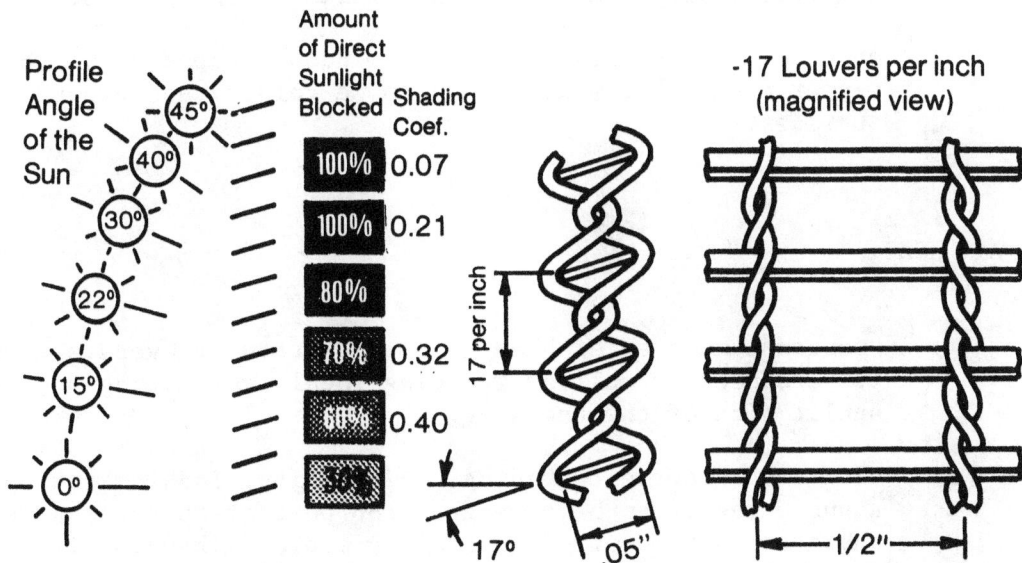

Figure 15.   Sun Penetration Through A Sun Screen vs. Sun Angle

2) The potential heat burden of sunlight penetrating a sunscreen
    early in the morning or late in the afternoon is small,
    because the sun's intensity is diminished at these times.
    This is due to the increased distance through the atmosphere
    which the light must travel. (68,Pennington,p88) This is
    illustrated below.

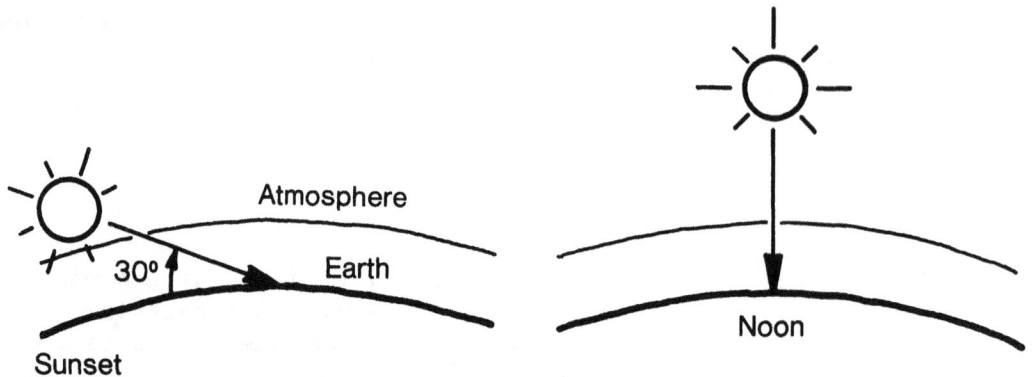

Figure 16.   Distance Through Atmosphere vs. Sun Angle

3) An external sun screen installed close to a window creates a
    layer of virtually still air, thus preserving the layer of air
    at the surface of the glass. This benefit results in a reduc-
    tion in the winter U-value of a single glazed window with a
    sun screen from 1.13 to 0.85 (74,ASHRAE,p407). (See STRATEGY:
    Windbreak for discussion of air films at the surface of glass.)

4) An external sun screen blocks a window's exposure to the
    winter night sky, which tends to be colder than ground
    surfaces.

ADVANTAGES:

1) Reduced summer solar heat gain which reduces average interior
    temperatures as well as reducing local overheating on the
    sunlit side of the building.

2) View and concurrent shading. Visibility, looking out through
    a sun screen, can be as high as 86 percent as through an
    unprotected window. (75, Koolshade,p16) The visibility is
    greatest when the sight line is parallel to the angle.

3) Daytime privacy without eliminating the view out.

4) Unobstructed night-time surveillance of building interior through windows with sun screens.

5) Reduced direct and reflected glare. Sun screens block light reflected from the ground or adjacent buildings more effectively than awnings, roof overhangs, or wall fins.

6) Reduced heat loss in winter due to protection of air film at the surface of the window, and due to reduced heat radiation to the cold night sky.

7) Protection of glass from projectiles thrown by vandals. (Replacement of the sun screen may cost more than replacement of the glass, but likelihood of injury from breakage is greatly reduced.)

8) Prevention of insect intrusion. Conventional insect screening is commonly a mesh of 16 x 18 per inch. One type of sun screen is equivalent to a mesh of 23 x 2 per inch and, hence, is likely to be suitable for insect screening.

9) May provide solar heating in the future. An experimental unit is under development which incorporates a sun screen in an air space between two layers of glass. In the summer the air space is vented outdoors at the top and bottom, and in the winter it is vented indoors to take advantage of the solar heat build-ups.

DISADVANTAGES:

1) Reduced solar benefit in the winter when sun screens are left in place.

2) No night-time privacy. In the night, the direction of transparency reverses - the screen appears opaque to occupants while allowing passersby an uninhibited view in. For residences, this necessitates the installation of drapes or blinds for privacy.

3) Interference with outswing windows.

4) Interference with window washing.

5) Impeded egress in the event of fire.

AESTHETICS:

1) Louvered sun screens make windows appear blackened on the facade. This increases their visual impact as a design element.

2) Louvered sun screens somewhat darken the view out of a window.

3) Louvered sun screens striate the view. The direction of the visual striation depends on whether the horizontal elements or the vertical elements are thicker for a given viewing angle.

COSTS:

Prices of louvered sun screens constructed of aluminum horizontal slats held with twisted bronze wire and finished with a black heat absorbing coating are approximately:

$6.00 per square foot for large windows.
$6.50 for smaller windows (less than 4 foot x 5 foot).
$5.00 delivered price to "do it yourself homeowners"

The price of an expanded metal sun screen made from a sstamped single sheet is lower, at the expense of reduced transparency. The price ranges from $3 to $3.50 per square foot.

EXAMPLES:

1) The Department of Public Works for Buck County, PA, installed louvered sun screens on its administration building at a cost of $36,000. Annual savings in operating costs were estimated to be 35% or $16,283. Thus, the costs of the sun screens were amortized in just over two years. (76,ACHR,p.29)

2) Following the references are photographs of sun screen installations illustrate the daytime outside appearance, the effect on view out from the inside, and the night-time outside appearance of a sun shade. (76,Koolshade)

REFERENCES:

ACHR News, <u>Screening Windows from Sun Helps Cut Cooling Costs 35 Percent; Heating Costs 28 Percent</u>, Air Conditioning, Heating, and Refrigeration News, Birmingham, MI, Oct. 29, 1976; Oct. 29, 1973.

ASHRAE, <u>ASHRAE Handbook of Fundamentals</u>, ASHRAE, Inc., New York, NY, 1974.

Halleck, Edward, personal communications. Construction Specialties, Cranford, NJ, Jan. 5, 1976.

Kaiser, <u>Louvered Aluminum for Solar Control</u>, Kaiser Aluminum, Oakland, CA, 1956.

Koolshade, <u>Shades Save Power</u>, Koolshade Corp., Solana Beach, CA, 1976.

Pennington, C. W., <u>How Louvered Sun Screens Cut Cooling, Heating Loads</u>, Heating, Piping, and Air Conditioning, Reinhold Publishing Co., Stamford, CT, Dec. 1968.

Kool Shade Corp., 722 Geneviewe St., Solana Beach, CA 92075

2.2   EXTERIOR ROLL BLINDS/Shading, Insulation

STRATEGY:

>   Install exterior roll blinds to provide sun shading in the
>   summer and to reduce winter heat flow.

PHENOMENA:

1)   Horizontal slats on a roller at the head of a window can be
        lowered to provide an opaque barrier to the summer sun,
        blocking both direct and diffuse sunlight. One manufacturer
        reports up to a 35-percent reduction in air conditioning costs
        with the use of roll blinds. (76,Frowein,p1) The following
        figure illustrates the installation of an exterior roll blind.

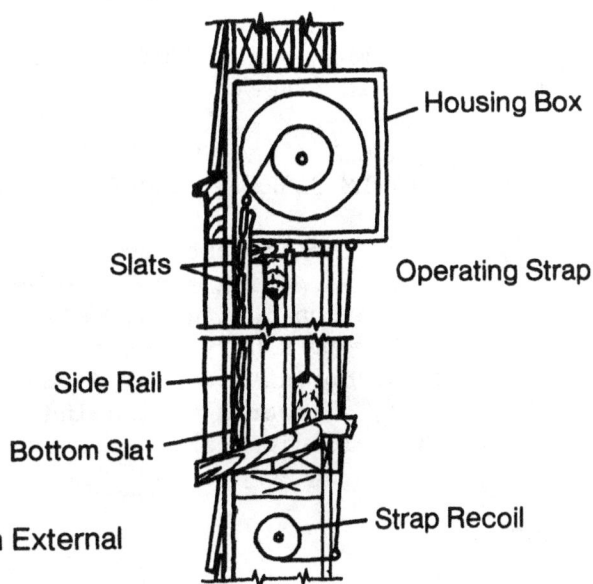

Figure 17.   Installation of an External
             Roll Blind

2)   When the slats are in the lowered position but not yet resting
        one on top of the next, horizontal slots between the slats
        permit air to circulate through the blind.

3)   If the roll blind tracks are hinged, the entire lowered roll
        blind can be projected out from the window during the day to
        provide natural ventilation concurrent with shading.

4)   Light colored roll blinds more effectively keep rooms cooler, because the blind reflects incident sunlight and remins cooler than would be the case with a dark-colored blind.

5)   During the winter, when the blind is lowered and the slats rest one on top of another, the resulting layer of air trapped between the blind and the window acts as insulation. Exterior roll blinds provide the most effective insulating air space of all the exterior strategies reported, because of the tight joints between the slats and the seal provided at the top and sides. One manufacturer reports the following U-values for roll blinds in combination with various types of windows. (76,Pease,p2)

GLASS TYPE	SEASON	GLASS ALONE	GLASS + 1/2 x 2" ROLL BLIND SLATS	GLASS + 1/8 x 1 3/8" ROLL BLIND SLATS
SINGLE	WINTER	1.13	0.405	0.568
	SUMMER	1.06	0.395	0.550
DOUBLE (1/2" AIR SPACE)	WINTER	0.58	0.301	0.384
	SUMMER	0.56	0.297	0.376
SINGLE + STORM SASH	WINTER	0.56	0.297	0.376
	SUMMER	0.54	0.290	0.366

5)   Since the coldest hours of the day occur during the hours of darkness, using a roll shade during the night provides increased insulation during the period of greatest potential heat loss. In New York City, for example, 70 percent of the degree-days occur during hours of darkness. (76,Claridge,p.57) Furthermore, a lowered roll blind obstructs a window's exposure to the cold night sky, further reducing heat loss.

ADVANTAGES:

1)   Shading during summer days with the roll blind lowered, and unimpeded ventilation at night with the roll blind rolled up into the head of the window.

2)   Insulation from winter heat losses with the roll blind lowered at night, and unimpeded solar gain through the window with the roll blind raised during the day.

3) Independent control of the shading of each individual window, and ability to partially shade a window in the case of the sun striking only part of the window, or where it is desirable to shade only a portion of a room.

4) Protection of glass from vandalism and wind storms when shade lowered.

5) Deterrence to burglars entering through windows when shade lowered.

6) Impediment to spread of flames out a window and up the side of the building. (Metal Roll Shades)

7) Reduction in noise transmission with blinds lowered. One manufacturer reports a noise reduction from 100 DBA to 60 DBA or a STC value of 40. (Mar 76,Frowein,p.1)

DISADVANTAGES:

1) Slats cannot be tilted venetian blind fashion to provide shade and view.

2) Delayed egress in the event of a fire.

3) Maintenance required for operating hardware, and for slats if wooden.

4) Limited to maximum single span of 12 feet for vertical windows (less for sloped windows), and a maximum height of approximately 10 feet. Large size exterior roll shades should be motor operated.

AESTHETICS:

1) In the raised mode, the slats are rolled into a concealed pocket in the head of the window.

2) In the lowered mode, a roll blind has the appearance of an opaque panel with horizontal grroves on the facade. They are available in colors; in wood, vinyl, or aluminum; and in different slat sizes.

3)   In the lowered position where the slats do not yet come in
     contact with each other, the effect from the interior is a
     series of horizontal slits of light. When the blind is locked
     down, it appears as a solid opaque panel blocking almost all
     daylight. In this lowered mode, if the frame assembly is
     hinged, it can be cranked out like an awning. Ground-reflected
     light will illuminate the ceiling of the room. If the window
     is above the second floor, the activity on the street or
     grounds below can be viewed through the resulting opening.

COSTS:

     The cost of a roll blind per square foot decreases with larger
     units, because much of the cost of manufacturing the unit is
     in the roller assembly. The following prices are for a unit
     without installation: (76,Sinnock)

APPLICATION	WIDTH	HEIGHT	$/FT^2	TOTAL $
WINDOW	3'-0" x	3"-6"	10.12	106.26
SL. GL. DOOR	6'-0" x	7'-0"	7.20	302.40
SL. GL. DOOR	11"-6" x	7'-0"	6.00	483.00

EXAMPLES:

1)   Roll blinds have been used extensively in Europe for several
     decades. Today they are used on approximately 25% of all
     European residential buildings and commercial high-rise
     buildings.

     In the United States, designers have only in the last three or
     four years begun to use roll blinds.

     Following the references are two photographs illustrating a
     roll blind viewed from the inside and viewed from the outside.
     (76,Frowein)

2)   Because of their effectiveness in reducing winter heat loss
     and summer heat gain, roll blind devices have been included in
     two recent demonstration energy conserving houses: the Zero
     Energy House built by the Technical University of Denmark, and
     the NASA Technology Utilization House in Hampton, Virginia.

REFERENCES:

Claridge, David, Window Management and Energy Savings, Efficient Use of Energy in Buildings, LBL 4411, Lawrence Berkley Laboratories, Berkley, CA, 1976.

Frowein, Onno J., personal communication, ROTO International, Essex, CN, March 15, 1976.

Frowein, Onno J., Letter, ROTO International, Essex, CN, March 17, 1976.

Harboe, Knud Peter and Koch, Soren, Zero Energy House - a model in 1:1, Rapport No. 114, Insitute for Husbygning, Polytekniske Laereanstalt, Lyngby, Denmark, 1976.

NASA, NASA Technology Utilization House, Tech. Brief LAR-12134 NASA Langley Research Center, Hampton, VA, 23665, 1976.

Pease, Amrol Exterior Rolling Shutters Conserve Fuel, Provide Privacy and Security, Pease Company, New Castle, IN, 1976

Sinock, Pomeroy, Telephone conversation, Pease Co., New Castle IN, Dec. 14, 1976.

Steuff, Horst, Konstruktive Moeglichkeiten im Rolladenbau, Bundesverband Deutscher Rolladenherstelle, 516 Deuren, Alte Jeulicher St. 105, Postfach 210, West Germany, 1971 (in German).

Roto International, P.O. Box 73, Essex, CT 06426

Roto International, P.O. Box 73, Essex, CT 06426

2.3 ARCHITECTURAL PROJECTIONS/Shading

STRATEGY:

Design architectural projections to shade windows from summer
sun.

PHENOMENA:

1)   Horizontal or vertical plane(s) projecting out in front
     of or above a window can be designed to intercept the
     summer sun, admit much if not all the winter sun, and
     allow a view out. If the plane projects far enough from
     the building a single projection may be sufficient as in
     the case of generous roof overhangs or windows recessed
     deeply between vertical fins. Alternatively, more modest
     projections can be equally effective in shading but they
     must be more closely spaced.

2)   East and west-facing windows are more effectively shaded
     by vertical projecting planes, south-facing windows are
     more effectively shaded by horizontal projecting planes.

3)   For shading effectiveness the color of the projection
     should be dark to reduce the light reflected off the
     projection and through the window. The light absorbed by
     this dark color is converted to heat and then dissipated
     to the outside air without becoming an air conditioning
     load. A separating gap between the shading device and
     the window is important to provide free circulation of
     the air to insure this heat dissipation.

4)   For daylighting effectiveness the underside of a horizontal
     projection should be light colored to reflect indirect
     ground reflected light into the room.

5)   The further south a building is located, the more important
     shading east and west-facing windows becomes and the less
     important shading south-facing windows becomes. This is
     due to the high position of the summer sun in southern
     latitude with the resulting decrease in direct sunlight
     transmitted by the south-facing windows. (See Strategy,
     Orientation to Sun)

ADVANTAGES:

1) Reduced summer solar heat gain. If air can circulate between the shading device and the window, and the window is completely shaded from direct sunlight the solar heat gain of the window can be reduced by as much as 80 percent. ('67,ASHRAE,p.485)

2) Reduced glare on work surfaces adjacent to windows.

3) Reduced winter heat loss to the sky.

4) Possible shelter from winter winds with corresponding reduction in heat conduction losses at the surface of the glass and reduction in infiltration.

5) Simplified window washing where wide horizontal projections can secondarily serve as a working platform.

DISADVANTAGES:

1) Possible impediment to window washing in the case of narrow closely spaced shading planes.

2) Possible impediment to fire egress in the case of narrow closely spaced shading planes.

3) Obstruction of view.

4) Increased maintenance. Horizontal planes will collect dirt, bird droppings, and ice.

AESTHETICS:

1) A single overhead projection with a light colored underside will cut off the blue light from the sky but admit the red or green light of ground reflected light.

2) Vertical projections from either side of the window narrow the peripheral view from the window.

3) Deeply recessed windows afford a framed view to the outside with sight lines quickly cut off when the viewer moves away from the center of the window.

4) Closely spaced horizontal or vertical planes may begin to dominate the view out of a window and in any case change the scale of the window. The proportion of the space divided by the shading planes becomes as important as the overall window proportion in determining the aesthetic effect of the fenestration.

5) Horizontal projections can provide a sense of security in the instance of floor to ceiling windows in tall buildings.

COSTS:

Generally, custom design and therefore not subject to cost generalizations.

EXAMPLES:

1) The ASHRAE Handbook of Fundamentals (73,ASHRAE,p409) contains a table which gives the distance a horizontal projection must extend out from the wall to shade an area up to 10 feet below the projection from April through September. Projections are given for eight orientations and for latitudes from 24 to 56°N in 8 degree increments. The projections are calculated for each hour.

2) The following photographs illustrate:

a) Closely spaced horizontal shading projections.

b) Deeply recessed windows providing vertical and horizontal shading projections.

c) Closely spaced vertical shading projections.

REFERENCES:

ASHRAE, ASHRAE Handbook of Fundamentals, American Society of Heating, Refrigeration and Air Conditioning Engineers, Inc., New York, 1967 and 1973.

Olgyay, Victor and Aladar, Solar Control and Shading Devices, Princeton Univ. Press, Princeton, N. J., 1957.

Belinda Collins, Sensory Env. Sect. NBS, Washington, D.C. 20234

2-18

Belinda Collins, Sensory Env. Sect., NBS, Washington, D.C.   20234

Belinda Collins, Sensory Env. Sect., NBS, Washington, D.C.   20234

2.4 EXTERIOR SHUTTERS/Shading, Insulation, Air Tightness

STRATEGY:

Install louvered, operable shutters outside a window to
provide shade in the summer during the day and reduce heat
loss and infiltration in the winter during the night.

PHENOMENA:

1) An exterior shading device effectively blocking all
   direct sunlight can reduce solar heat gain through a
   window up to 80 percent. (undated,NBS,p2) The shading
   performance of closed exterior shutters depends upon how
   well the heat absorbed by the shade itself is dissipated
   to the outside air. Operable louvers adjusted to block
   the sun but let air circulate improve the shutter's
   ability to keep out heat. Similarly, light colored
   shutters which reflect much of the sunlight rather than
   absorb it are more effective. (Actual shading coefficients quantifying performance could not be located.)

2) Heat loss through a window with closed shutters is reduced
   because the air space between the shutter and the glass
   provides additional resistance to the flow of heat to the
   outside. How effective the shutter is in reducing heat
   loss depends upon the air tightness of the space between
   the shutter and the glass. A shutter with pivoting
   louvers which can be closed is beneficial to this end.
   However, even louvers fixed in an open configuration
   reduce heat loss through the window by substantially
   sheltering the insulating film of air at the outer surface
   of the glass from the scouring action of the wind, and by
   reducing infiltration through window cracks.

ADVANTAGES:

1) Reduced solar heat gain in the summer.

2) Management by the occupant on an individual basis to
   permit control of shading, light level within the room,
   and view out.

3)   Protection from rain penetration through windows opened
     for ventilation.

4)   Reduced night heat loss in the winter.

5)   Protection of windows from storm damage, vandalism, or
     intrusion.

6)   Privacy.

DISADVANTAGES:

1)   Operation requires reaching outside the window which
     necessitates insect screen or storm sash being mounted
     inside the window and being openable.

2)   Subject to wind damage if not secured properly.

AESTHETICS:

Shutters are considered by many residential designers and
apparently the home buying public, mandatory dressing of
windows. Regretfully, the fact that shutters can serve
valuable energy conserving functions as well as cosmetic
functions has been forgotten. If their potential benefit is
again realized, perhaps operable shutters will again be an
option to more home buyers in the future.

COSTS:

Primed, wooden shutters with fixed open louvers are commonly
available at lumber yards. A sample of prices in Washington,
D. C. is given below. Hardware to make exterior shutters
operable is not commonly available as such but can be devised
simply from gate hinges.

      - 15 x 39 inches ----------------$12.00 per pair

      -    x 47 -----------------------$15.00

      -    x 51 -----------------------$19.00

      -    x 55 -----------------------$25.00

Wooden shutters with adjustable-tilt slats could not be found
to obtain sample costs. The common interior shutter for
interior use is not heavy enough construction to be recom-
mended for exterior use.

Prefinished aluminum shutters with fixed open slats are also
available. The following are a sample of retail prices in the
Miami area. (See EXAMPLES for a description of the types
given below.)

- Bahama $4.25 per sq. ft.

- Sarasota $4.50

- Rolling $5.00

- Side-hinged $6.00

Vinyl shutters are frequently used to dress windows. These
are not appropriate for use as operable shutters if they are
molded to be seen only from one side or are molded with
continuous simulated louvers with no open space between the
slats.

EXAMPLES:

1)   Several types of operable shutters are common in the
      southern regions of the U. S. The following are examples
      of various modes of operation and associated generic
      names:

Bahama Shutters

Sarasota Shutters

Rolling Shutters

Side-hinged Shutters

Figure 18. Types of Shutters

2) The photographs following the references illustrate
    installations of Bahama and Sarasota types of operable
    shutters as seen from the outside and the quality of
    light penetrating a closed rolling shutter as seen from
    the inside. (76,Wilk)

REFERENCES:

NBS, "Home Energy Saving Tips". National Bureau of Standards,
    Washington, D. C., undated.

Wilk, James A., personal communication, Willard Shutter Co. Inc., Miami
    Fla., Nov. 5, 1976.

Willard Shutter Co., 4420 NW, 35 Court, Miami, FL 33142

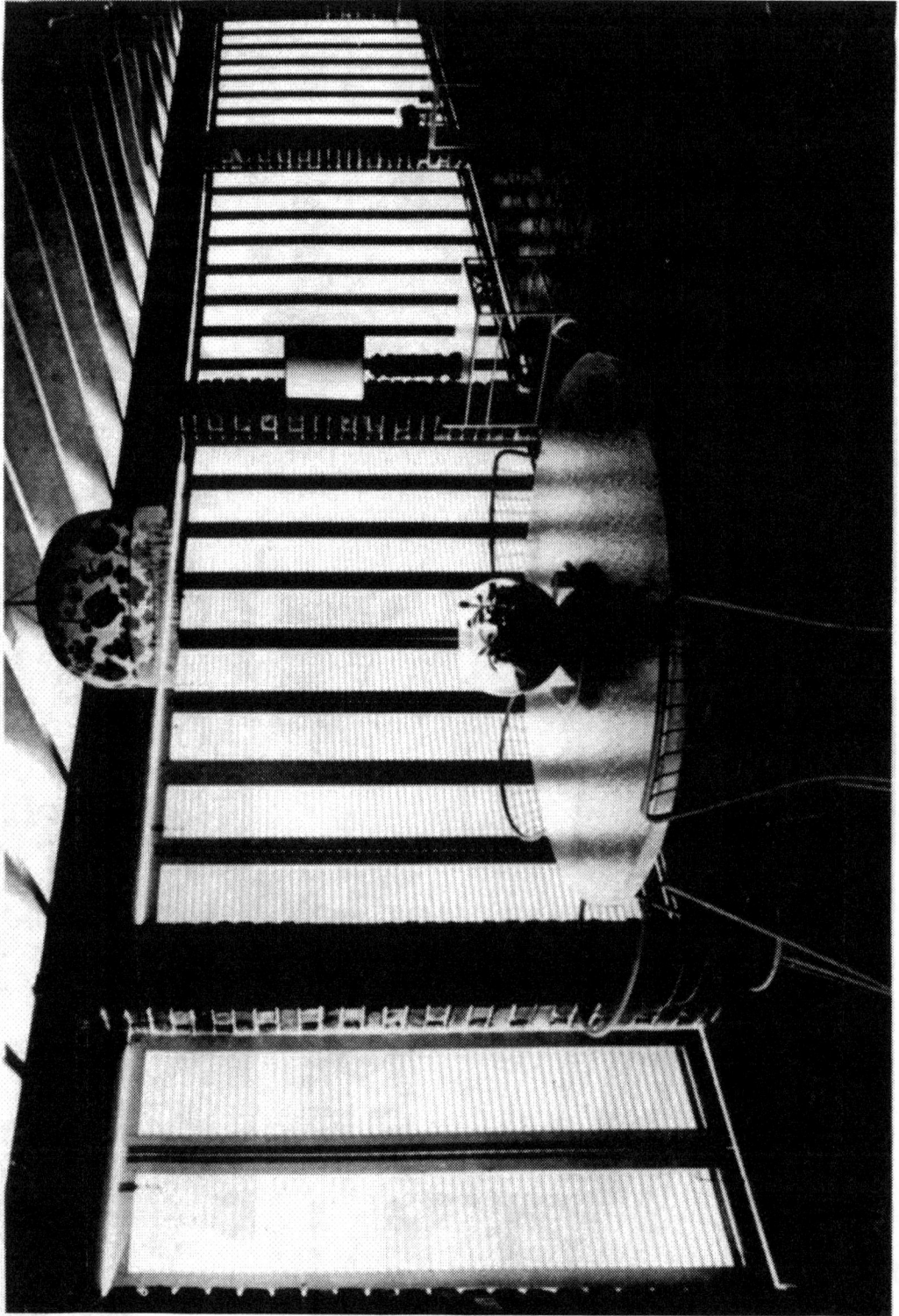

Willard Shutter Co., 4420 NW, 35 Court, Miami, FL 33142

2.5 AWNINGS/Shading

STRATEGY:

Install an awning with provision for air to circulate between it and the window to provide shade without heat build-up.

PHENOMENA:

1) How well an awning shades a window is dependent on how opaque the material is to both direct sunlight and diffuse light from the sky. The following table gives the transmittances (amount of light penetrating/total incident light) of several common awning constructions: (58,Ozisik,p463)

SOLAR TRANSMITTANCES OF AWNING MATERIALS & CONSTRUCTIONS

Material	Direct transmittance	Diffuse transmittance
Canvas	0.0%	0.0%
Plastic	0.25	0.15
Aluminum[1] (separated slats)	0.0	0.20

Note:
1. Source: (76,Stolz) See the figure at the end of the strategy for an illustration of an aluminum awning.

2) The surfaces of the awning exposed to the sun should be a light color to minimize the amount of sunlight absorbed. Sunlight absorbed by the awning raise its temperature. Much of this heat may be transferred to the window in two ways: by radiation, and by raising the temperature of the air between the awning and the window. Light colored awning materials are therefore more effective because they stay cooler and transfer less heat to the window. For example, a white canvas awning or a slatted, white aluminum awning reflects between 70 and 91 percent of the sunlight depending on how clean it is (dirt absorbs light). By comparison, a dark green canvas awning reflects only 21 percent, and a dark green plastic awning reflects 27 percent of the sunlight. (58,Ozisik,p466)

3) The heat from sunlight absorbed by a dark-colored fabric will build-up under the awning and be transferred to the window unless air is permitted to circulate behind the awning. Fabric awnings are commonly installed with a narrow continuous gap between the top of the awning and the wall to prevent hot air from being trapped under the awning. Slatted aluminum awnings inherently provide air circulation by virtue of the gaps (ranging from 1/4 inch to 3/4 inch) between the horizontal slats.

4) In order for an awning to be effective, it must be designed to provide adequate coverage of the window area for the specific orientation of the window. A south-facing window requires only a minimal horizontal projection to be completely shaded all summer, all day. An east or west-facing window needs an awning which extends down a substantial percentage of the window height in order to provide protection from the low sun angles of early morning or late afternoon. In addition, the sides of the awning should be closed to prevent sunlight from angling in behind the awning on south-facing windows. The following table illustrates the sunlit area of glass beneath awnings of various configurations in Cleveland, Ohio. (58,Ozisik,p472)

PERCENTAGE OF GLASS AREA SUNLIT FOR VARIOUS AWNING CONFIGURATIONS

ORIENTATION	SOLAR TIME[1]		WITH SIDE PANELS AWNING DROP[2]			OPEN SIDES AWNING DROP[2]		
			0.65	0.60	0.55	0.65	0.60	0.55
SOUTH	8 a.m.	4 p.m.	4	8 %	13%	7%	14%	22%
	9 a.m.	3 p.m.	2	4	7	5	11	18
	10 a.m.	2 p.m.		2	5	2	5	7
EAST	7 a.m.	5 p.m.	7	13	22	7	13	22
WEST	8 a.m.	4 p.m.		7	15		7	15

Note:

1) Solar time approximates clock time half-way across a time zone. Clock time at the western extreme of the time zone will be approximately one-half hour earlier than solar time, and clock time at the eastern extreme of a time zone will be approximately one-half hour later than solar time.

2)   The awning drop equals the distance the awning
     extends down the window divided by the total window
     height.

5)   The net effectiveness of an awning in reducing the summer
     solar heat gain of windows is given below for a design
     day representing August 1 at 40 degrees latitude. Heat
     gain is totaled for the period from 8 a.m. to 4 p.m. for
     the south exposure and noon to 5 p.m. for the west
     exposure. The awnings have a 70 percent drop and pro-
     vision to vent air at the top. A dark foreground is
     assumed. For a light foreground the heat gain could be
     as much as approximately twice the amount shown due to
     light reflecting up beneath the awning. (58,Ozisik,p474)

HEAT GAIN THROUGH SINGLE GLAZED WINDOWS WITH AWNINGS

ORIENTATION OF WINDOW	TYPE OF AWNING	HEAT GAIN PER 100 SQ FT. GLASS SURFACE, BTU/DAY	HEAT EXCLUDED BY THE AWNING	
			BTU/DAY	PERCENT REDUCTION
SOUTH	No awning	62,200	0	0
	White canvas awning	22,500	39700	64
	Dark green canvas awning	27,700	34500	55
	Dark green plastic awning	35,600	26600	43
WEST	No awning	84,200	0	0
	White canvas awning	19,500	64700	77
	Dark green canvas awning	23,900	60300	72
	Dark green plastic awning	34,800	49400	59

ADVANTAGES:

1)   Reduced summer solar heat gain by up to 55 to 65 percent
     on south-facing windows and 72 to 77 percent on west-
     facing windows. (58, Ozisik,p475)

2)   Reduced glare.

3)   Rain protection for windows opened to provide ventilation.

4)   Unobstructed view out in a downward direction.

5)   Removable in winter to let sunlight in and prolong the
     life of fabric awnings.

DISADVANTAGES:

1) Subject to wind damage.

2) Periodic replacement of fabric due to weathering deterioration. (Canvas: 4 to 6 years, vinyl coated canvas and
   plastic: 6 to 8 years)

3) Reduced effectiveness if ground surfaces and/or adjacent
   vertical surfaces are highly reflective.

4) Horizontal view out partially obstructed, view of sky
   largely or completely obstructed.

5) Rainwater run-off from large awnings can cause splash
   problems on the ground.

AESTHETICS:

1) Awnings are available in a variety of colors and patterns
   They can be a bright, cheerful addition to an otherwise
   drab facade.

2) Awnings will drastically darken the building interior by
   eliminating the sun and bright sky as two sources of
   illumination. Any sunlight which does penetrate at the
   bottom of the window area does not project any depth into
   a room.

COSTS:

In the Washington, D. C. area the following is a sample of the
installed cost of an awning covering a 3 ft. by 5 ft. high
residential window:

MATERIAL	CONFIGURATION	COST OF FABRIC AND FRAME	COST OF FABRIC REPLACEMENT
Painted Canvas	No side panels	$60	$40
	w/side panels	$70	$50
Vinyl-coated canvas	10 percent more than painted canvas		
or vinyl-coated dacron			
Acrylic treated acrilan	20 percent more than painted canvas		
Enameled aluminum	solid panels	$100	
	open slatted	$120	

EXAMPLES:

1) The following example illustrates first cost savings
    possible with awnings on residential windows. Calcula-
    tions are for a room with 400 square feet of floor area,
    and two normal-sized, unshaded windows facing west. If
    awnings are installed over the two windows the reduced
    heat load permits the use of a 3/4 HP motor to drive the
    a/c compressor instead of a 1 HP motor. The smaller size
    also permits the use of the standard 110 volt electrical
    service rather than separate wiring providing 220 or 230
    volts. Thus, the use of awnings saves $60 to $100 in the
    purchase price of the air conditioner and $50 to $100 in
    installation costs for separate wiring. (Ogden,p4)

2) An example of first-cost savings possible with canvas
    awnings on a commercial application is the newly constructed
    Administration Building at North East Missouri State
    University. By using an opaque, acrylic awning over the
    windows, less expensive gray, tinted glass could be substi-
    tuted for reflective glass and the size of the air condition
    ing system could be reduced due to the reduced solar load.
    Operating costs of the heating system as well as the air
    conditioning system are expected to be lower (due to the
    winter solar gain admitted when the awnings are removed)
    in comparison to reflective glass which rejects sun year-
    round. (76,Keller)

3) Operating cost savings can be calculated for an example case using the table listed under the PHENOMENA (number 5). White canvas awnings are installed on the west side of a house in New York City. (40°-46' lat.) There are six west-facing windows totaling 100 square feet. The awnings would reduce the cooling load by 64,700 BTUs per day. Assuming an air conditioning system consumes one KWH to remove 6826 BTUs, 9.5 KWH would be saved. At $0.04/KWH the savings amount to $0.38. This represents August 1, a day when the air conditioning load is likely to be greater than a day in the beginning or end of the air conditioning season. However, the amount of daily solar radiation (the heat source awnings reduce) in August is actually less than the average for the period of May through September. Therefore, the savings calculated for August 1 are a conservative estimate of daily savings possible during the air conditioning season.

4) Following the REFERENCES is an example of a roller awning used on an office building, (75,Avery,p1) and a slatted, aluminum awning used on a residence. (Alcan,p2)

5) The following are examples of common awning configurations

Roller Awning
(self-storing)

Hood Awning

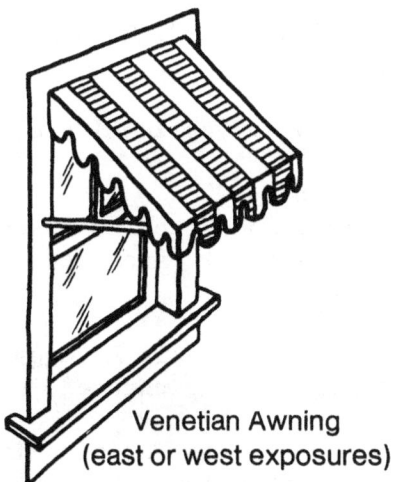

Venetian Awning
(east or west exposures)

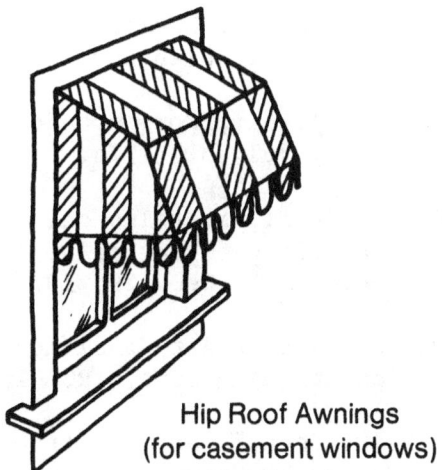

Hip Roof Awnings
(for casement windows)

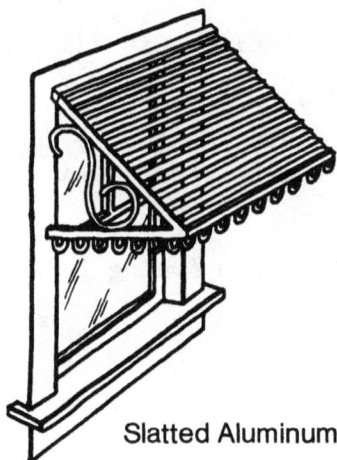

Slatted Aluminum

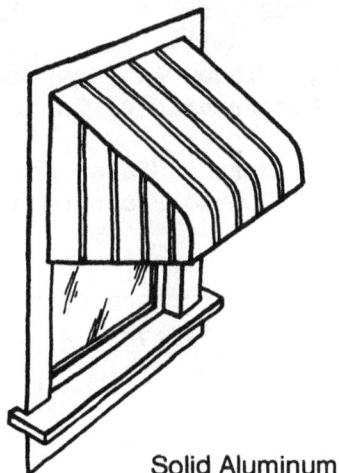

Solid Aluminum

Figure 19.   Types of Awnings

REFERENCES:

Alcan, "Flexalum Awnings", Alcan Building Products, Cleveland, Ohio, undated.

Avery, "Queensland Sunblind", J. Avery & Co. Ltd., 82-90 Queensland Road, Holloway, England., February, 1975.

Buckingham, Donald. Telephone conversations, Washington Shade and Awning Co., Gaithersburg, Md., Dec. 27, 1976.

CPAI, "The Utility and Distinction of Design in Canvas", Canvas Products Assoc. Int., Saint Paul, Minn., 1964.

Glen Raven, "Sunbrella Outdoor Decorating Guide", Glen Raven Cotton Mills Inc., Glen Raven, N. C., undated.

Grehan, Arthur, Correspondence, American Canvas Institute, Memphis, Tenn., July 31, 1975.

Keller, William, telephone conversation, John Steffen Assoc., consulting engineers to Ittner and Bauersox, Architects, St. Louis, Mo., Dec. 29, 1976.

Ogden, J. B., "Air Conditioners Will Help You Sell Awnings", Air Conditioning Dept., RCA Inc., Chicago, Ill., undated.

Ozisik, Necati, and Schutrum, L. F., "Heat Gain Through Windows Shaded by Canvas Awnings", ASHRAE Transactions, Vol. 64 ASHRAE Inc., New York, N. Y., 1958.

P.G & E., "Window Awnings Save Energy", Pacific Gas and Electric Co., San Francisco, Ca., undated.

Schultz, Kenneth, "Solar Shading with Canvas Awnings", Canvas Products Review, Canvas Products Assoc. & Int., St. Paul, Minn., March 1965.

Stolz, Ivan, telephone conversation, Aluminum Awning Industries, Stockton, Cal., Dec. 27, 1976.

J. Avery & Co., 82-90 Queensland Rd., Holloway N7 7AW, England

Alcan Building Products, P.O. Box 511, Warren, Ohio 44482 Mr. Aiknis

FRAME

FRAME VENTILATORS
WEATHERSTRIPPING
THERMAL BREAK
TYPE OF OPERATION
WINDOW TILT
SIZE, ASPECT RATIO

3

3. FRAME

The window frame can enhance or detract from the beneficial energy attributes of a window. The material from which it is constructed can be insulating or highly conductive of precious winter heat and prone to condensation problems. Weatherstripping between operable sash and the frame can substantially impede infiltration through joint cracks at the perimeter of the window. The perimeter can be kept minimal by adjusting the proportion of the window. Temperate breezes can be captured and their flow into a building directed by the choice of operating window type. Finally, the winter solar heat gain can be increased and summer solar gain rejected by merely tilting the window frame.

3.1 FRAME VENTILATORS/Ventilation

STRATEGY:

Specify window frames with a provision for controlled admittance of outside air through the frame section into the building interior.

PHENOMENA:

Small openings can be incorporated in the head or sill section of the frame to admit fresh air without rain or insect penetration. A weatherstripped shutter can provide tight closure when ventilation is not desired.

ADVANTAGES:

1) Occupant control of the ventilation.

2) Ventilation at window where security or cost preclude operable sash.

DISADVANTAGES:

Lack of centrally controlled admittance of outside air.

AESTHETICS:

A deeper head or sill frame section to accommodate the ventilation openings is required.

COST:

Including through the frame ventilation increases the frame cost between 15 and 20 percent depending upon the unit selected.

EXAMPLES:

Following the references are examples of frame ventilators.

REFERENCES:

Kawneer, "Kawneer Has Just Re-Invented the Window", Kawneer Architectural
    Products, Niles, Mich., 1976.

Roto International, "Unitas-Dauerluftung" Roto International, Essex,
    Conn. 1975.

Wausau Metals Corp., "Aluminum Windows and Curtain Wall", Wausau Metals
    Corp., Wausau, Wisc., Jan. 1975.

Kawneer Architectural Products, Country Club Rd., Harrisonburg, VA 22801

Roto International, P.O. Box 73, Essex, CT 06426

Roto International, P.O. Box 73, Essex, CT 06426

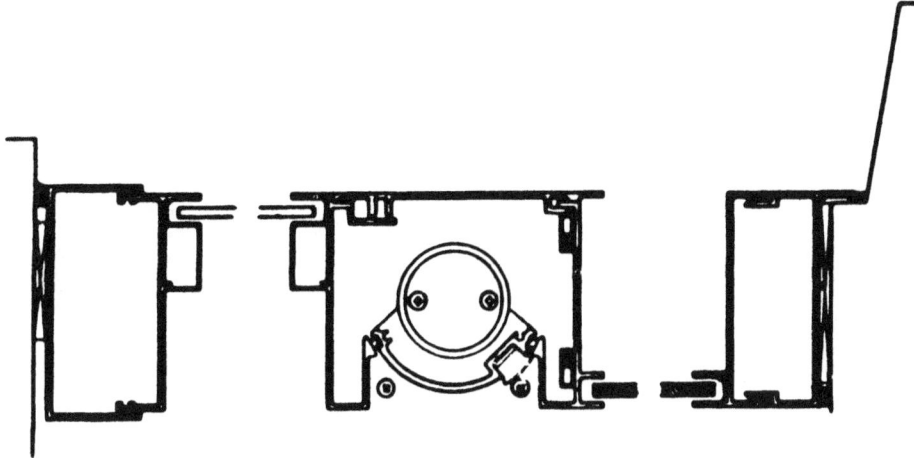

Wausau Metals Corp., 1415 West Sheet, Wausau, WI 54401

3.2 WEATHERSTRIPPING/Air Tightness

STRATEGY:

Install weatherstripping to reduce air leakage through windows.

PHENOMENA:

1) Infiltration is one of the primary ways that energy is
    lost through windows. Every three feet of edge of
    operable sash may lose as much energy as one square foot
    of glass. (74,Professional Builders,p154).

2) Air leakage through cracks only occurs when there is a
    difference in air pressure between the inside and outside
    of the building. There are two fundamental causes of
    this air pressure difference: 1) outside wind induced
    pressure and or inside mechanical system induced pressure.
    2) air density difference due to inside and outside air
    temperature difference. These two causes can tend to
    cancel each other or can be compounding.

3) The air tightness of a window depends upon the initial
    size of the crack between the frame and sash necessary
    for the sash to be movable and to accomodate fabrication
    tolerances; and upon the change in the crack size with
    aging due to general wear, distortions of the frame due
    to external stresses transferred from the building, and
    shrinking or warping of the components. Air leakage
    through the perimeter joints of operable sash can be
    effectively reduced with weatherstripping because of the
    ability of weatherstripping to accommodate changing joint
    sizes.

4) ASHRAE estimates the effectiveness of weatherstripping
    for various types of windows as follows: (65,ASHRAE,p459)

CUBIC FEET OF AIR INFILTRATION PER FOOT OF CRACK, HOUR

	Wind Speed (mph)				
Wood double hung (average fit, unlocked)	5	10	15	20	25
non-weatherstripped	7	21	39	59	80
weatherstripped	4	13	24	36	49
average reduction	38 percent reduction due to weatherstripping				
(poor fit, unlocked)					
non-weatherstripped	27	69	111	154	199
weatherstripped	6	19	34	51	71
average reduction	70 percent reduction due to weatherstripping				
Metal Double hung (unlocked)					
non-weatherstripped	20	47	74	104	137
weatherstripped	6	19	32	46	60
average reduction	60 percent reduction due to weatherstripping				

ADVANTAGES:

1) Reduced infiltration of outside air on the windward side of a building and reduced loss of conditioned air on the leeward side of a building.

2) Elimination of uncomfortable drafts.

3) Increased resistance to water and snow prentration.

4) Improved sound insulation.

DISADVANTAGES:

1) Deteriorates from physical aging or wearing.

2) Cannot compensate for gross frame distortions.

EXAMPLES:

1) An analysis was performed on a hypothetical five bedroom
bungalow with 998 square feet of floor area, 12 windows
with a combined area of 221 sq. ft., two doors with a
combined area of 37 sq. ft., and 4 inches of insulation
in the walls and ceiling. The windows were average
fitted and had rib-type metal weatherstripping. The
interior temperature was to be maintained at 70°F. The
following table summarizes the calculated reduction in
winter heating costs due to weatherstripping the windows
of the hypothetical bungalow located in various cities.
The costs have been recalculated by the authors for
natural gas at $0.21 per 100 cubic feet (Nov 76 price in
Washington, D. C.), 1000 Btu/cubic foot, and a furnace
efficiency of 80 percent. (52,Lund,p4)

FUEL COST SAVINGS FROM WEATHERSTRIPPING

City	Degree Days	Fuel Cost due to Infiltration			Total Fuel Cost	
		Non W.S.	W.S.	Savings	Non W.S.	W.S.
Washington, D. C.	4,561	55.23	19.40	35.83	149.72	113.89
New York City	5,280	63.93	22.47	41.46	173.30	131.85
Chicago, Ill.	6,282	76.06	26.72	49.34	206.20	156.86
Minneapolis, Minn.	7,966	96.46	33.89	62.57	261.50	198.92
Grand Forks, N. D.	9,871	119.53	41.99	77.54	324.03	246.49

W.S. = Weatherstripped windows

non-W.S. = non-weatherstripped windows

2) The following section through a double hung window illustrates
one manufacturer's weatherstripping method.

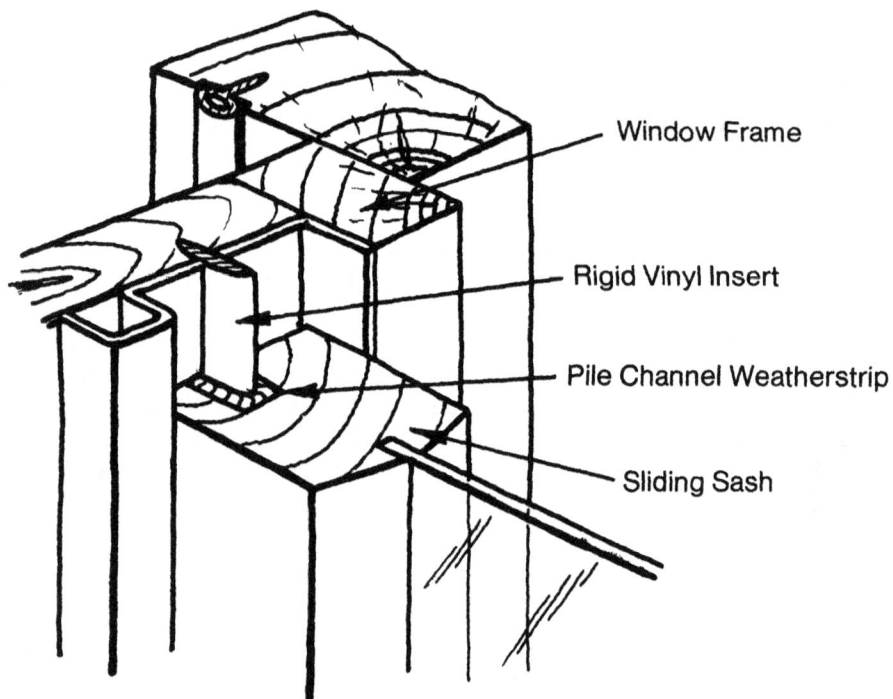

Figure 20. Cut-away Section of a Weatherstripped Window

REFERENCES:

ASHRAE, ASHRAE Guide and Data Book, American Society of Heating,
    Refrigeration and Air-Conditioning Engineers, New York, 1965.

Lund, M. S., Peterson, W. T. Air Infiltration Through Weatherstripped
    and Non-Weatherstripped Windows, Bulletin No. 35, University of
    Minnesota, Minneapolis, Minn. 1952.

Professional Builder, "Stop Housing's Biggest Energy Drain", Professional
    Builder, Chicago, Ill., April 1974.

Tamura, G. T., "Measurement of Air Leakage Characteristics of
    House Enclosures", ASHRAE Transactions No. 2329, ASHRAE Inc.,
    New York, N. Y.

3.3 THERMAL BREAK/Insulation

STRATEGY:

Provide a thermal break in the path of heat flow through metal window frames to reduce winter heat loss and summer heat gain.

PHENOMENA:

1)   Aluminum conducts heat 1,770 times better than wood and therefore offers little resistance to unwanted flow of heat. (75,Kern,p.47)   This inherent disadvantage of metal window frames can be alleviated by thermally separating the inside of the frame from the outside of the frame.   There are presently two methods of providing a thermal separation:

   a)   pouring poly-urethane in a slot in the metal frame; then after it has bonded and set, sawing away the metal bridging the slot.

   b)   providing two separate frames linked together with a rigid vinyl insert.

2)   The effectiveness of the thermal break depends upon the insulating value of the material used and the thickness of the material in the path of the heat flow.

   An aluminum frame with a good thermal has a U-value similar to insulating glass (U=0.58) and performs substantially better than an aluminum frame with no thermal break as shown below: (76, Kolbishop)

	Frame U-Value
2 inch thick aluminum frame - no break	U = 1.18
2 inch thick aluminum frame - with break	U = 0.60

3)   The benefit of providing a thermal break in the frame is proportional to the amount of frame area. The frame area can be as high as 20 percent of the total window area. In such a case, if the window is glazed with insulating glass, it is important that the frame not provide a "short circuit" for the heat flow. A thermal break in the frame prevents this.

ADVANTAGES:

1) Reduction in winter heat loss and summer heat gain through the window frame.

2) Elimination of condensate or ice forming on the frame except in the most extreme conditions.

3) Elimination of wall deterioration due to run-off from condensate.

DISADVANTAGES:

1) Possible degradation of the thermal break from sun exposure if the frame section does not provide protection.

2) Imposition of stress on the thermal break if it is bonded to both sides of long sections of aluminum subject to wide inside/outside temperature differences and extreme summer/ winter temperature ranges.

   For example, the temperature of the outer frame, if it is anodized a dark color, can range from 160°F (71°C) when exposed to sunlight in the summer to below zero (-17.7°C) on cold winter nights. The frame inside the thermal break will stay much closer to room temperature, especially if the window has reflective or heat absorbing insulating glass. The wide seasonal temperature range of the outside section will cause it to expand and contract while the inner section changes relatively little in length. (76,Hetman) This phenomena requires the use of a resilient material for the thermal break and a good bonding agent. (76,Roehm)

3) Increased aluminum cross section size required for structural integrity if the thermal break allows free slippage between the two frame parts.

AESTHETIC CONSEQUENCES:

1) The frame sections can be detailed to obscure the thermal break from view or the thermal break can be colored to blend with the frame.

2) A more massive frame section may be necessary to accommodate the thermal break and still have adequate structural integrity if the thermal break does not bond the two sections together.

COSTS:

The cost of installing window frames with a thermal break depends
upon the complexity of the frame. For simple fixed glass frame
sections, a thermal break adds approximately 10 percent to the
material cost of the frame. For openable windows, a thermal break
will cost considerably more because there are more frame components
and they must withstand the stresses of operation.

EXAMPLES:

1) A manufacturer has calculated the energy and resulting cost
    savings for installing windows with a thermal break in a new
    office building in Lincoln, Nebraska. (6,671 heating degree
    days, 1,282 cooling degree days) All windows were to be
    double glazed and non-operable. The building had 1,350 sq.
    ft. of inside window frame area. Oil heating with an effi-
    ciency of 70 percent was assumed. Based upon an improvement
    in the frame U-value from 1.18 to 0.60 it was calculated
    that 1,270 gals oil could be saved. At $0.40 per gallon
    this amounts to $508 saved per heating season. Additional
    savings are expected from reduced air conditioning costs.
    The thermal break interrupts the conduction from the sun-
    heated outside surface of the frame.   The additional cost
    of providing window frames with a thermal break was $1080.
    (76,KOBISHOP)

2) The following illustrations show how poured poly-urethane
    thermal breaks can be included in window frame sections.

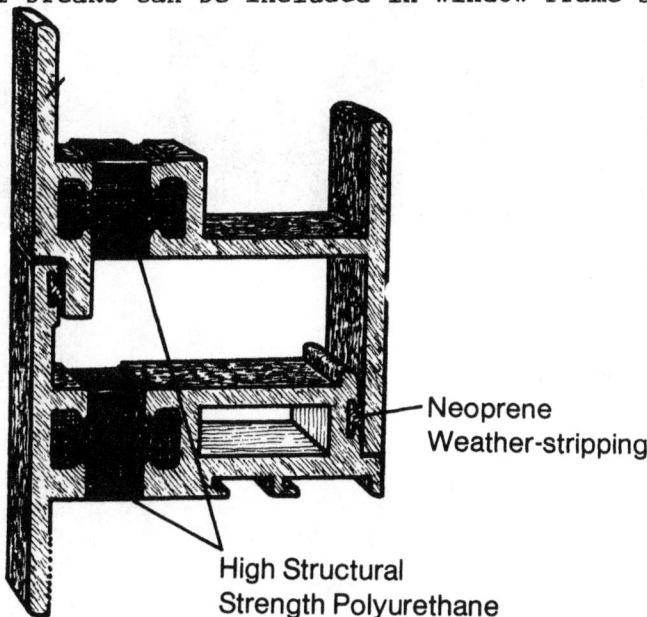

Neoprene
Weather-stripping

High Structural
Strength Polyurethane

3) The following photograph illustrates the effectiveness of a thermal break in reducing heat loss through the frame. A piece of dry ice is placed in contact with the outside face of two window frames. The frame section to the right has no thermal break and ices-up, the frame section on the left contains a thermal break and does not ice-up on the inside. (76,Devac,p5)

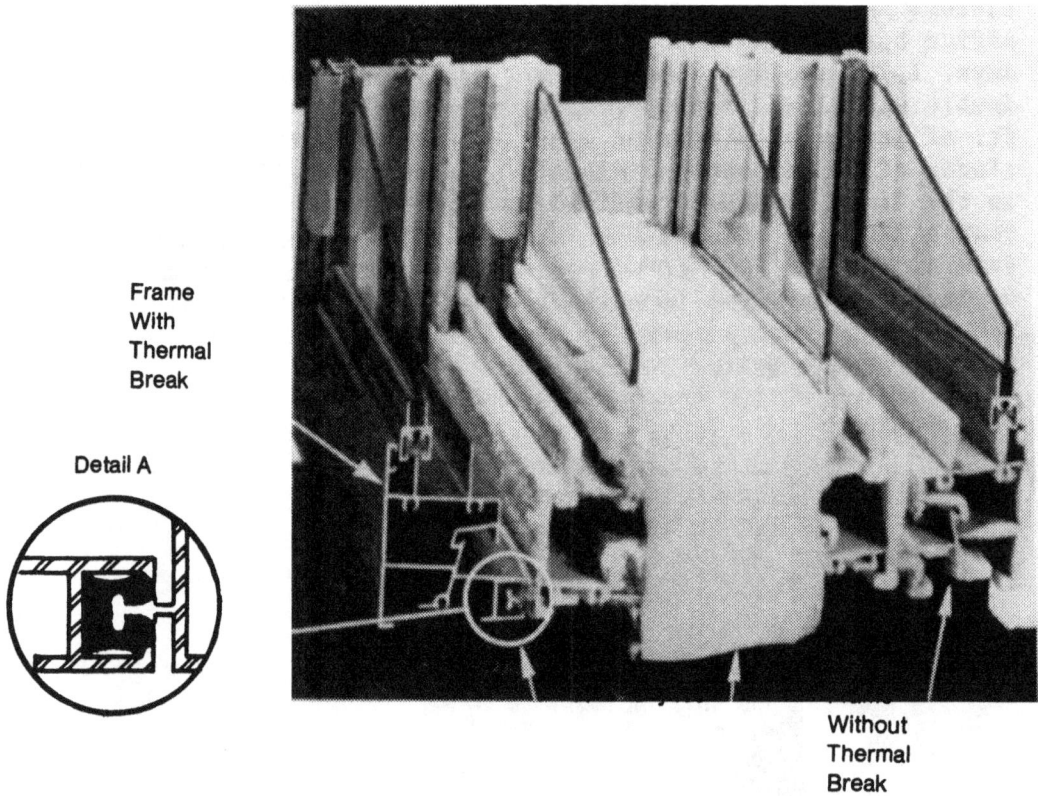

Frame
With
Thermal
Break

Detail A

Without
Thermal
Break

Devac, Inc., 10130 State Highway 55, Minneapolis, Minn. 55441

3-14

REFERENCES:

Hetman, Frank, telephone call by author, Devac, Inc., Minneapolis,
  Minn., Dec. 2, 1976.

Kern, Ken, The Owner Built House. Charles Scribner Sons, New York,
  1975.

Kobishop, Peter, correspondence addressed to author. Wausau Metals
  Corp., Wausau, Wisconsin. Nov. 24, 1976.

Roehm, John, telephone call by author, Architectural Aluminum Manufacturing Assoc. Chicago, Ill., Dec. 3, 1976.

3.4 TYPE OF OPERATION/Ventilation, Air Tightness

STRATEGY:

Select a type of operating window considering its ability to draw in outside breezes and direct incoming air.

PHENOMENA:

1) Outward projecting casement windows can scoop in and exhaust air when the wind is parallel to the wall. (55,Jones,p4)

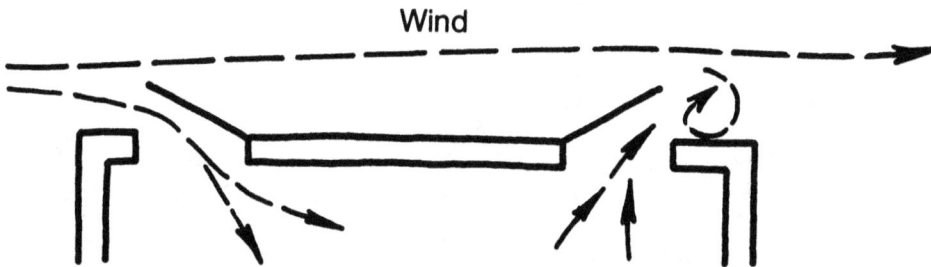

Wind

Figure 21.   Plan View of Casement Windows

2) Top hinged, bottom hinged, center pivoting, and jalousie windows can angle the incoming air stream upward in the plane of the sash. (73,Olgyay,p111) This helps relieve the tendency of hot air to stagnate near the ceiling.

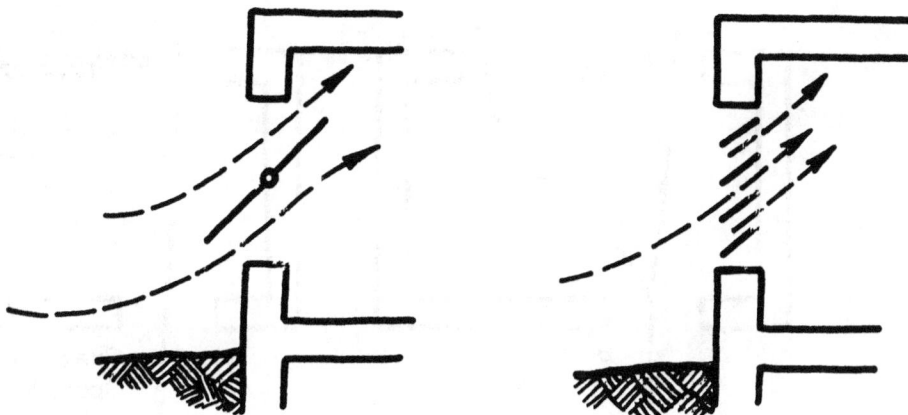

Figure 22.   Section Through Horizontal Pivoting Windows

3) Double hung windows provide slight ventilation, even on windless days, due to the difference in density between warm and cool air. When the outside air is cooler than room temperature, warm, less dense room air exits out the top window opening while cool, more dense outside air is drawn in through the bottom window opening replacing the exiting warm air. If the outside air is warmer than room air the process reverses. Because this effect increases with increasing vertical separation of the top and bottom openings, tall, narrow windows are more effective ventilators. (74,Grandjean, p211)

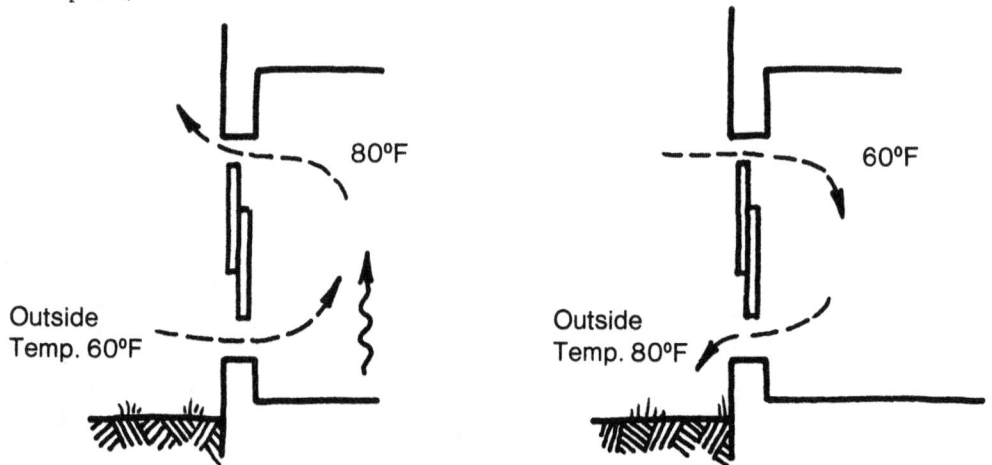

Figure 23.   Air Circulation Through Double-Hung
Windows Shown in Section

4) Horizontally sliding windows with both halves operable may provide more circulation of air in what might otherwise be dead air spaces. Also, the option of opening one side, the other side, or both affords more flexibility for furnishing a room and more options for the room occupant to direct the air flow.

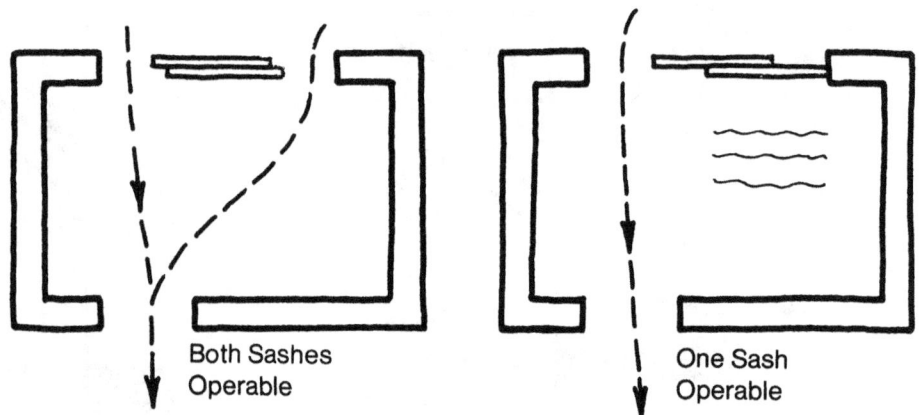

Figure 24.   Plan View of Air Movement Through
Horizonal Sliding Windows Shown in Plan

ADVANTAGES:

1) More effective overall ventilation of a room.

2) Ventilation during rain possible with out-swing awning, in-swing hopper, and jalousie windows.

3) 100 percent of window area is openable with pivoting and hinging type windows compared to a maximum of 50 percent with vertically or horizontally sliding windows.

4) Greater ease in window washing possible when outside glass surfaces can be reached through opened sash, when sash can be pivoted nearly 180°, or when sash can be lifted out of tracks.

5) Occupant control. Option to have greater variation of room temperature and ventilation than likely with mechanical system alone.

DISADVANTAGES:

1) Possible wind damage when opened, hinged, or pivoting sash catch gusts.

2) More parts to require maintenance with vertically or horizontally pivoting sash compared to sliding sash.

3) Possible entry for burglars by removing glass slats of jalousie windows.

4) Interference with inside draperies, roll shades, or venetian blinds with in-swing windows, and interference with outside sun screens, roll blinds, or canvas awnings with out-swinging windows.

The following table summarizes several of the above advantages and disadvantages of window operating types:

	WINDOW TYPES										
ADVANTAGES	horizontal sliding	double hung	double hung (reversed)	casement (out)	casement (in)	pivoted (vertical)	pivoted (horizontal)	top hinged (out)	bottom hinged (in)	fixed sash	jalousie
provides 100% vent opening				X	X	X	X	X	X		X
diverts inflowing air upward							X		X		X
will deflect drafts				X	X	X	X		X		
offers rain protection while partly open							X	X	X		X
screen and storm sash easy to install	X	X	X	X	X				X		
easy to wash with proper hardware	X		X		X	X	X		X		
DISADVANTAGES											
only 50% of area openable	X	X	X								
does not protect from rain when open	X	X	X	X	X	X					
inconvenient operation when over an obstruction	X	X	X				X	X			
presents a hazard if vent low and close to walkways				X		X	X	X			
hard to wash		X		X						X	X
interferes with furniture, drapes, blinds, etc.					X	X	X		X		
screens-storm windows difficult to provide	X					X	X				
sash has to be removed for washing		X		X					X	X	

AESTHETICS:

1) Subdivision of a window to provide operable as well as fixed portions reduces the scale of the fenestration and may reduce dimensions to a more human scale.

2) With a large facade, occupant discretion in having the windows closed, partially open, or completely open will provide a constantly changing pattern to the composition of the fenestration.

3) An open window, unlike a ventilation register delivering "processed" air, allows a sense of contact with the outdoors, both visually, acoustically, and olefactorily.

4) Horizontal window meeting rails must be designed with eye level, sight-lines, and view considered in order that annoying view obstruction be avoided.

5) Too much subdivision of the glass area can distract from an attractive view, as with jalousie windows.

COST:

The additional cost of selecting the correct but more expensive type of window to ventilate a room versus selecting the least expensive type of window is small relative to the total building cost. Providing double hung versus single hung results in minimal if any additional costs. Often the track and separate sash are already existant, only the operating hardware need be provided. Casement windows are more expensive than horizontally or vertically sliding windows but afford complete opening of the window area. Jalousie are more expensive than casement windows but afford 100 percent openable area plus rain protection.

EXAMPLES:

An unusual type of operating window is available which can be hinged at the side like an in-swing casement, or by shifting a lever control, hinged at the bottom like an in-swing hopper window.

When the sash is hinged at the bottom and the heat register or radiator is located below the window the rising hot air will be directed into the room while being mixed with a controlled amount of outside fresh air. Drafts are minimal in this situation. When the sash is hinged from the side 100 percent of the window area is available for air circulation. Washing is facilitated by simply swinging the window into the room in the side hinged mode.

The in-swing allows the installation of external storm windows, insect screens, solar screens, or awnings. However, it does interfere with interior window accessories, e.g. draperies, venetian blinds, or shades as well as limiting the placement of furniture near the window. (75,Architects Journal,p488)

REFERENCES:

Architects Journal, "Windows Tech. Study No. 1, Hanging and Operation, Ironmongery", Architects Journal, Architectural Press Ltd., London, Eng., Sept. 3, 1975.

Becket and Godfrey, Windows, Van Nostrand, Reinhold Co., N. Y., 1974.

Grandjean, E., Ergonomics of the Home, Halsted Press, N. Y., 1974.

Jones, R. A. et.al., "Selecting Windows", Small Homes Council, Urbana, Ill., 1955.

Olgyay, Victor, Design with Climate, Princeton University Press, Princeton, N. J., 1973.

"Roto-Tilt and Turn Windows", Roto International, Essex, Conn., 1976.

3.5 WINDOW TILT/Solar Heating

STRATEGY:

Design windows tilted slightly towards the ground to reduce summer solar heat gain without appreciably affecting winter solar heat gain.

PHENOMENA:

1) Reflection of sunlight at the surface of glass varies considerably depending on the incident angle at which the light strikes the glass (i.e. the angle between the light ray and a perpendicular line from the surface of the glass as shown in the figure below). At incident angles

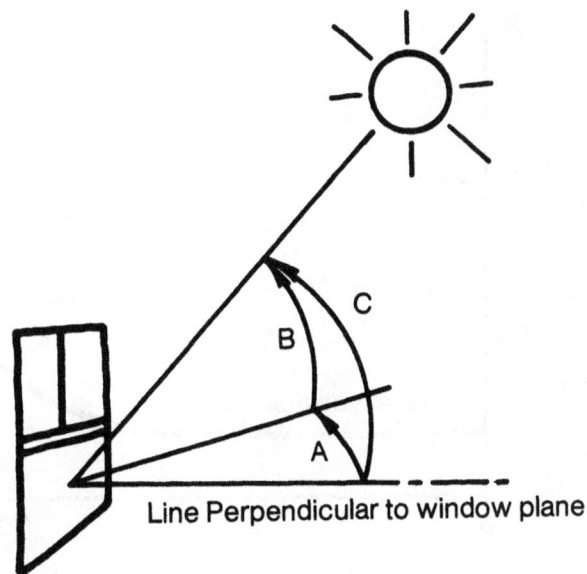

Line Perpendicular to window plane

A — Angle between sun and window in plan view

B — Angle between sun and ground plane (Altitude)

C — Incident angle

Figure 25.   Incident Angle Illustrated

less than 57° small changes in the incident angle have
little affect on the amounts of light transmitted or reflected. At incident angles greater than 57° the amount of
direct sunlight reflected increases at an increasing rate
and the amount of light transmitted decreases correspondingly
(77,Cellarosi) This is illustrated in the following graph.
(74,Yellot,p22)

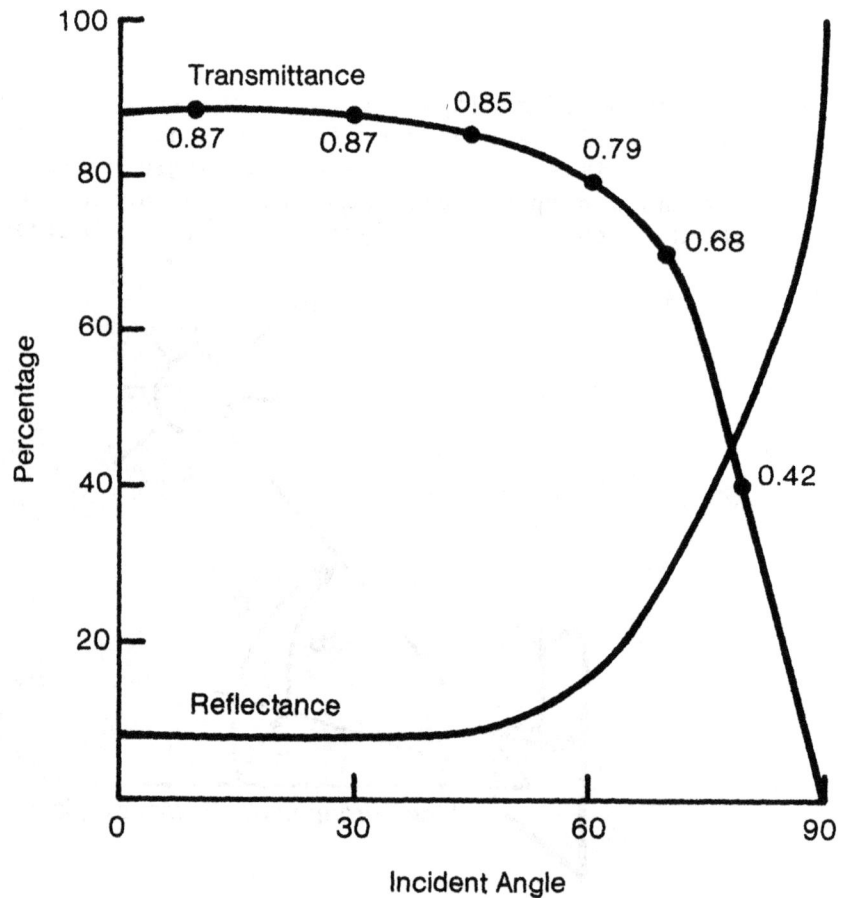

Figure 26.   Light Transmitted and Reflected by
             Glass vs. Incident Angle

2)   During the summer when the sun is high in the sky, the
incident angle of the sun on vertical glass is likely to be
well above 57°. For example, at 42° latitude at noon on
June 22 the incident angle of sunlight on a vertical plane
of glass facing south is 71.45°. This is so far in excess
of 57° that any increase in the incident angle, (i.e.
tilting the glass outward) will greatly increase the amount
of light reflected and reduce the solar heat gain correspondingly.

During the winter, when the sun is low in the sky, the
incident angle of sunlight on vertical glass is likely to be
less than 57°. For example, at 42° latitude, on Dec. 22 at
noon the incident angle between the sun and a south-facing
window is 24.55°. At this low sun angle, a slight tilt
downward to the glass will not appreciably decrease the
amount of sunlight transmitted in comparison with vertical
glass. Thus, the potential benefit of winter solar heat
gain is not appreciably decreased. This seasonal variation
is illustrated in the following figure.

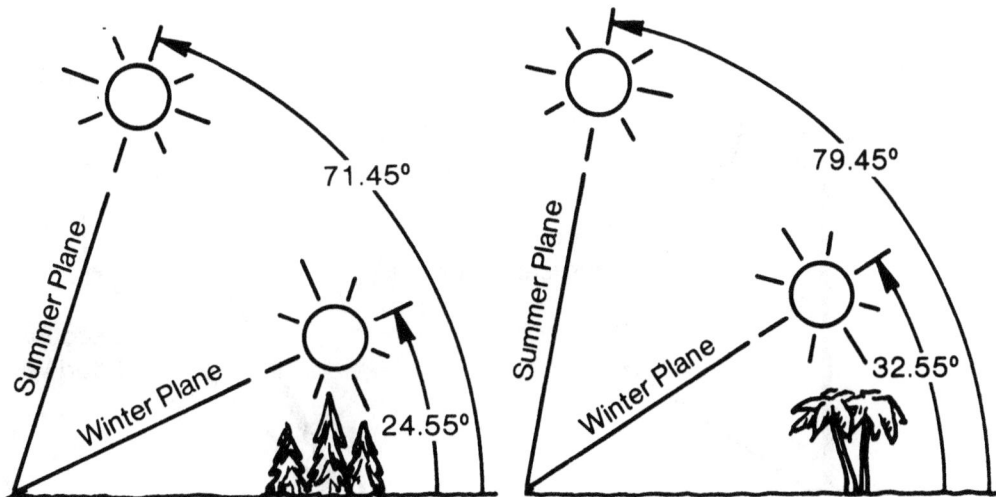

Altitude at Noon @ 42°N Lat. (Boston)	Altitude at Noon @ 34°N Lat. (Atlanta)
June 22 = 71.45°	June 22 = 79.45°
Dec 22 = 24.55°	Dec 22 = 32.55°

Figure 27.   Seasonal Sun Angle Variation For Two Cities

From the previous figure it can also be seen that closer to the equator (e.g., 34°N latitude) the sun is higher in the sky in both summer and winter. Thus, less tilt of the window is required to achieve the same reduction in summer sun transmission as is possible with more tilt in more northern latitudes.

3) The amount of summer sunlight transmitted through a tilted window is not only reduced because more of the light is reflected but it is still further reduced because the horizontal outward projection of the tilt also reduces the area of glass exposed to the sun. The geometry is the same as if there were a horizontal projection shading vertical glass. This shading effect becomes negligible in the winter when the sun is at a lower angle in the sky.

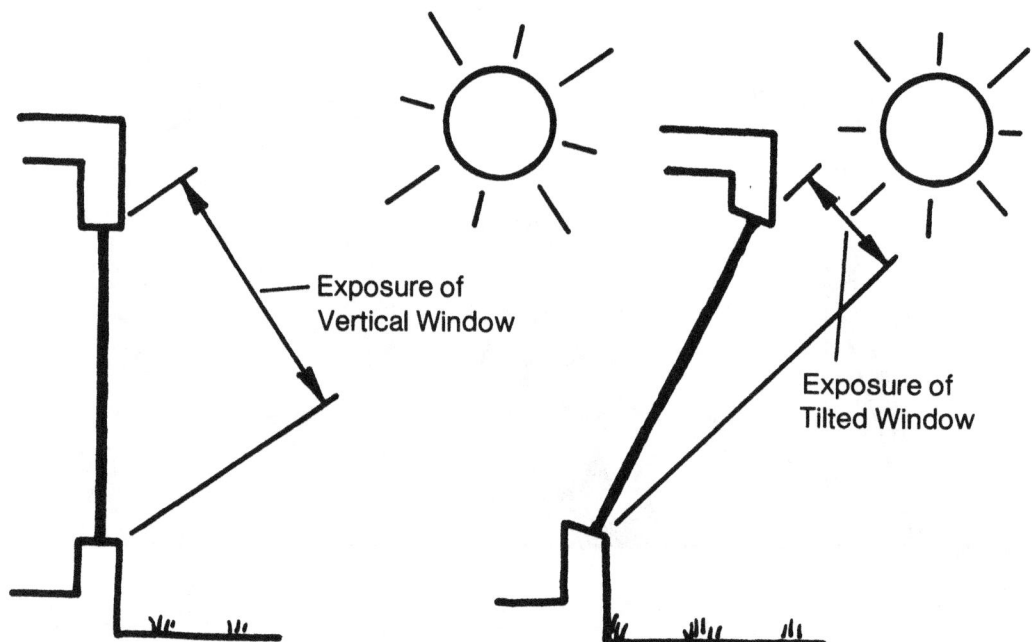

Figure 28. Glass Exposure to Sun Shown in Section View

ADVANTAGES:

1) Reduced summer solar heat gain without obstructing view out.

2) Little reduction of winter solar heat gain.

3) Reduction in exposure to cold winter night sky.

4) No management of the window required to achieve these effects.

5) Reduced glare.

6) Less prone to soiling from birds.

7) Protection of opened sash section from rain penetration.

DISADVANTAGES:

1) Increased exterior surface area for same floor area provides greater potential heat loss.

2) Not effective on east or west exposures when sun is low in the sky. (The more off of a true N.S.E.W. orientation a building is, the more effect tilting the glass will have on increasing the incident angle for the easterly or westerly exposures.)

3) Reduced effectiveness when ground surfaces are reflective

4) Dust more prone to collecting on inside surface of the glass.

5) Washing tilted windows may be more difficult than washing vertical windows.

6) Stronger roll blinds required for non-vertical windows because the slats, when not vertical, may tend to bow from their own weight.

AESTHETIC CONSEQUENCES:

1) During the day the projection of a building outward towards an observer may instill a sense of overbearing dominance. The effect during the night will be minimal.

2) The greater building perimeter at the top of the windows compared to the base of the windows results in a greater expanse of ceiling which may give a sense of spaciousness. However, there is no increase in floor area.

3) Installation of draperies will result in an odd relationship between the "hang" of the drapes and the slope of the glass. With venetian blinds, the slope may create operating difficulties.

COSTS:

Tilting the glass increases the area of glass for a given height of window opening. This will increase the material and the installation costs.

EXAMPLES:

1) The North Carolina Blue Cross and Blue Shield Headquarters in Orange County, N. C. is oriented with its long axis running east/west with sloping reflective glass on the north and south sides to reduce summer solar heat gain. Wind acceleration was a concern during the design stages but wind tunnel tests showed no adverse wind conditions. Air conditioning costs are said to be significantly reduced as a result of the shape of the building and use of reflective glazing. (71,P.A.,p129)

Libbey-Owens-Ford Co., 811 Madison Ave., Toledo, Ohio

2) In the Tempe Municipal Building of Tempe, Arizona the
    glass slopes at 45 degrees to similarly act as its own
    sunshade and reduce heat gain by reflecting sunlight due
    to the increased angle of incidence. Heat absorbing
    glass and draperies are used to further reduce heat gain
    The area immediately behind the glass is isolated from
    offices and the heat that does penetrate is carried away
    by the air handling system. The end result is that only
    18 percent of the available solar heat reaches the
    occupied areas of the building. (71,P.A.,p111)

City of Tempe, P.O. Box 5002, Tempe, Ariz. 85281

3) Dulles Airport Terminal Building in Fairfax County, VA is
    another example. The glazing tilts outward, consistant
    with the formal lines of the building, providing effective
    solar heat gain reduction. The long tilted sides are
    oriented facing south and north. East and west sides are
    vertical to facilitate future building expansion.
    Untinted single pane glazing is used without any internal
    draperies, shades or blinds.

Robert Wehrli, Arch. Research Sect., NBS, Washington, D.C. 20234

Robert Wehrli, Arch. Research Sect., NBS, Washington, D.C. 20234

REFERENCES:

American Society of Heating, Refrigeration and Air Conditioning Engineers, ASHRAE Handbook of Fundamentals, ASHRAE, New York, 1974.

Cellarosi, Mario, telephone conversation, National Bureau of Standards, Washington, D. C., March 29, 1977.

Latta, J. K., Walls, Windows and Roofs for the Canadian Climate, Division of Building Research, Ottawa, Canada, 1973.

P.A., "Shaped for Savings", Progressive Architecture, Reinhold Publishing Co., Inc., Stamford, Conn., Oct. 1971.

Stephenson, D. G. and Mitalas, G. P., "An Analog Evaluation of Methods for Controlling Solar Heat Gain Through Window", Research Paper No. 154, Division of Building Research, Ottawa, Canada, April 1962.

Yellot, John I., "Solar Radiation and Its Uses on Earth", Energy Primer, Protola Inst., Menlo Park, Ca., 1974.

3.6 SIZE, ASPECT RATIO / Air Tightness

STRATEGY:

   Proportion windows so they approach a square and use fewer but
   correspondingly larger window to minimize window perimeter,
   thereby reducing the potential for infiltration.

PHENOMENA:

   1)   Window infiltration occurs at three joints:

        a)   the perimeter joint between the frame and the wall.

        b)   the perimeter joint between the sash and the frame.

        c)   the perimeter joint between the glass and the sash.

        Insulating glass conduction losses are greatest at the
        perimeter where metal or glass edging bridges the in-
        sulating air space.

   2)   It is possible to decrease the perimeter of a given area
        of window merely by making it closer to a square in
        proportion. As can be seen in the following figure, less
        perimeter occurs for any given area the smaller the width
        to height ratio is. Also the smaller the area, the more
        pronounced this phenomena is.

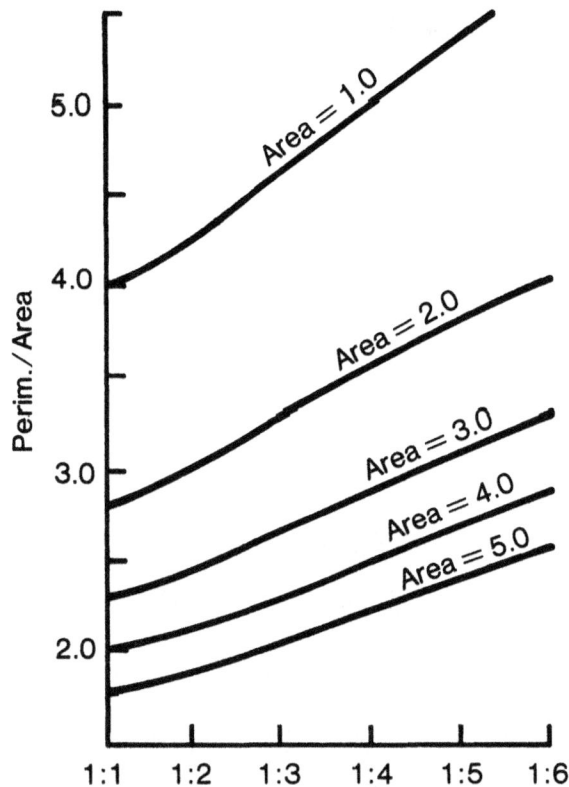

Figure 29. Aspect Ratio

3)   Regardless of the size of the area, the amount of perimeter for any given area increases at an increasing rate up to a width to height ratio of 1:3. Thereafter, the amount of perimeter increases at a decreasing rate. Thereafter, if non-energy criteria dictate a window narrower than a 1:3 ratio varying its slenderness has decreasing energy relevance.

4)   Using fewer but larger windows rather than more but smaller windows reduces the perimeter for a given window area. For example, two square windows each 3 ft. on a side provide a total area of 18 sq. ft. with a perimeter of 24 ft. A single square window 4.25 feet on a side provides the same area but the perimeter is only 17 ft.

ADVANTAGES:

1)   Reduced potential infiltration due to reduced perimeter.

2)   Reduced conducted heat loss through the edges of insulating glass.

3)   Less time required to clean a few large windows compared to many smaller windows.

4)   Reduced maintenance cost of painting and caulking.

5)   Reduced potential for water penetration.

DISADVANTAGES:

1)   Less uniform distribution of daylight with a few large
      windows compared to more windows placed at intervals.

2)   Fewer options for varying the sourced daylight and natural
      ventilation with fewer but larger windows.

3)   Increased physical effort required to operate large windows

4)   Increased cost for replacement of broken glass due to
      vandels or storm damage.

AESTHETIC CONSEQUENCES:

1)   Larger and square shaped windows may pose a problem if
      classical proportions are dictated in traditional design
      situations.

2)   Larger windows affect the scale of the building.

3)   Fewer but larger windows may result in a clearer delinea-
      tion of wall areas and window areas.

COSTS:

1)   The total delivered cost of fewer but larger windows is
      lower.

2)   The costs of framing and installing fewer but larger
      windows is apt to be less within certain ranges. One
      constraint is the limit of what a carpenter can handle.
      Secondly, the cost of framing the wall opening increases
      in discrete increments as the depth of the header increases
      in nominal increments.

EXAMPLES:

The figure below illustrates the proportions calculated in the table following the references for the two extremes of: area = 1, and area = 5.

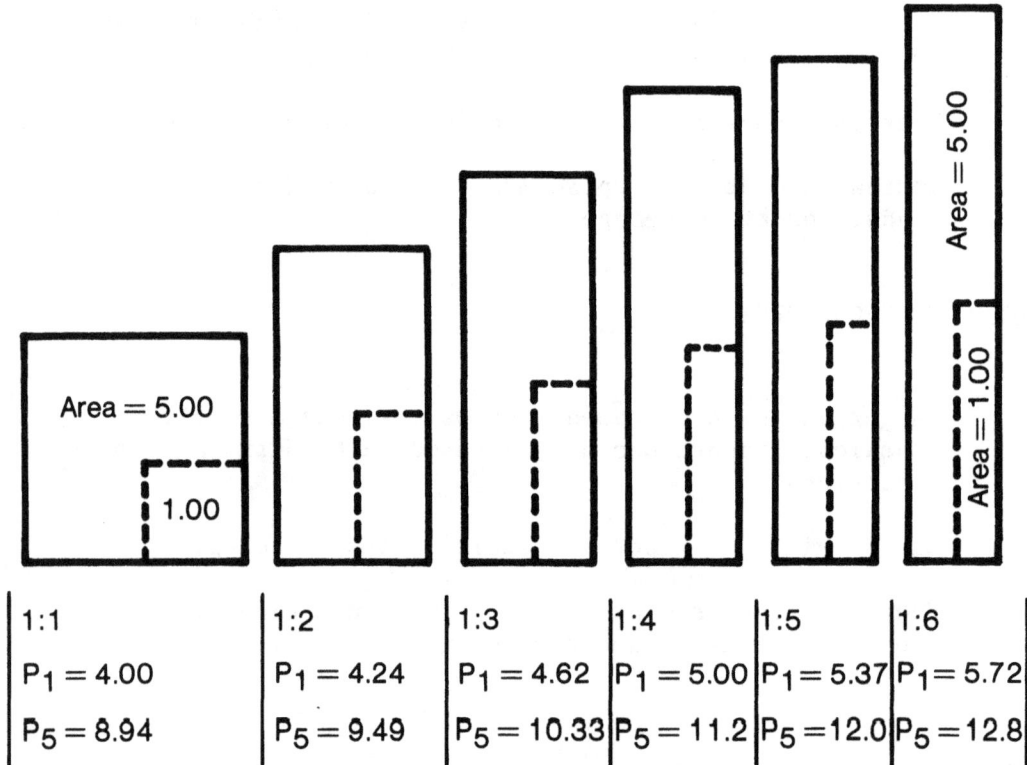

Figure 30. Aspect Ratios

1:1	1:2	1:3	1:4	1:5	1:6
$P_1 = 4.00$	$P_1 = 4.24$	$P_1 = 4.62$	$P_1 = 5.00$	$P_1 = 5.37$	$P_1 = 5.72$
$P_5 = 8.94$	$P_5 = 9.49$	$P_5 = 10.33$	$P_5 = 11.2$	$P_5 = 12.0$	$P_5 = 12.8$

REFERENCES:

Rehm, Ronald, meeting, Mathematics Division, National Bureau of Standards, Washington, D. C. Dec. 10, 1976.

The following table illustrates changes in perimeter and perimeter
to area ratios as a function of aspect ratios from 1 to 1
through 1 to 6 for areas of 1 through 5.

AREA = 1.0000

ASPECT	WIDTH	PER	ΔPER	PER/A	ΔPER/A
1 x 1.0	1.000	4.000		4.000	
1 x 2.0	.707	4.243	.111	4.243	.111
1 x 3.0	.577	4.619	.376	4.619	.376
1 x 4.0	.500	5.000	.381	5.000	.381
1 x 5.0	.447	5.367	.367	5.367	.367
1 x 6.0	.408	5.715	.349	5.715	.349

AREA = 2.0000

ASPECT	WIDTH	PER	ΔPER	PER/A	ΔPER/A
1 x 1.0	1.414	5.657		2.828	
1 x 2.0	1.000	6.000	.158	3.000	.079
1 x 3.0	.816	6.532	.532	3.266	.266
1 x 4.0	.707	7.071	.539	3.536	.270
1 x 5.0	.632	7.589	.518	3.795	.259
1 x 6.0	.577	8.083	.493	4.041	.247

AREA = 3.0000

ASPECT	WIDTH	PER	ΔPER	PER/A	ΔPER/A
1 x 1.0	1.732	6.928		2.309	
1 x 2.0	1.225	7.348	.193	2.449	.064
1 x 3.0	1.000	8.000	.652	2.667	.217
1 x 4.0	.866	8.660	.660	2.887	.220
1 x 5.0	.775	9.295	.635	3.098	.212
1 x 6.0	.707	9.899	.604	3.300	.201

AREA = 4.0000

ASPECT	WIDTH	PER	ΔPER	PER/A	ΔPER/A
1 x 1.0	2.000	8.000		2.000	
1 x 2.0	1.414	8.485	.223	2.121	.056
1 x 3.0	1.155	9.238	.752	2.309	.188
1 x 4.0	1.000	10.000	.762	2.500	.183
1 x 5.0	.894	10.733	.733	2.683	.183
1 x 6.0	.816	11.431	.698	2.858	.174

AREA = 5.0000

ASPECT	WIDTH	PER	ΔPER	PER/A	ΔPER/A
1 x 1.0	2.236	8.944		1.789	
1 x 2.0	1.581	9.487	.249	1.897	.050
1 x 3.0	1.291	10.328	.841	2.066	.168
1 x 4.0	1.118	11.180	.852	2.236	.170
1 x 5.0	1.000	12.000	.820	2.400	.164
1 x 6.0	.913	12.780	.780	2.556	.156

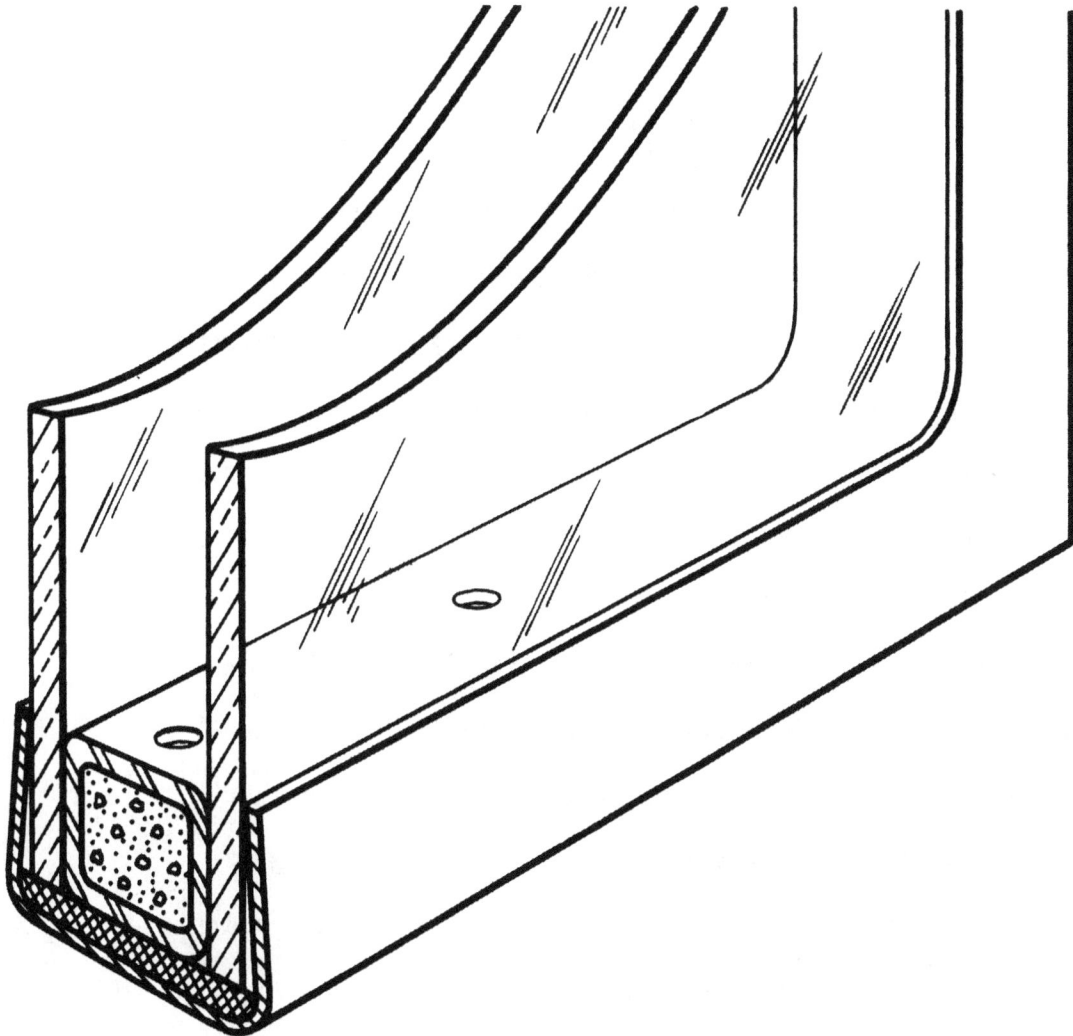

GLAZING

MULTIPLE GLAZING
HEAT-ABSORBING GLASS
APPLIED FILMS
REDUCED GLAZING
GLASS BLOCK
THRU-GLASS VENTILATORS

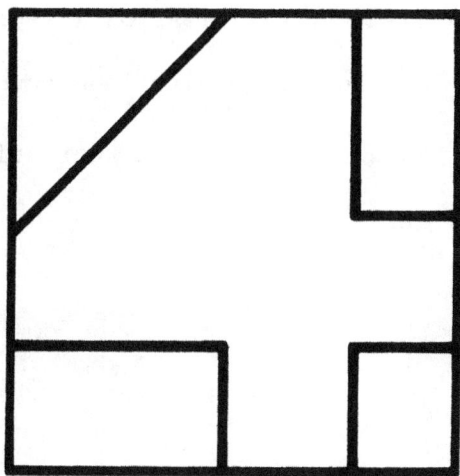

4. GLAZING

The type of glass installed in the window will determine the amount of sunlight transmitted into the building's interior and the amount of the building's heat conducted to the outside. These two factors will establish the heat gain or heat loss through glass for given interior and exterior climatic conditions. Insulating glass, multiple glazing (storm sash), and low-emissivity coatings are extremely effective at reducing the conducted heat flow through glass. Reflective or tinted glasses are capable of stopping much of the sunlight from penetrating into the building. In selecting glass, the orientation of the window, the length and severity of the seasons, and the heat gain from lighting equipment, and people must be considered. For example, clear double glass on a south exposure may be effective for locations with long severe winters, while reflective single glass may be appropriate for locations with long, hot summers.

4.1 MULTIPLE GLAZING/Insulation

STRATEGY:

Install insulating glass and/or storm windows to provide an insulating air space(s), reducing conducted heat losses.

PHENOMENA:

1) Glass is a good conductor of heat, be it indoor heat conducted outward in the winter, or outdoor heat conducted inward in the summer. The high conductivity of glass can be appreciated when compared to another material. For example, glass, as a material, conducts heat 9 times better than plywood. This rate of heat flow is so great that merely adding layers of glass in contact with each other is of negligible thermal benefit. However, if the layers of glass are separated by air spaces, the path of conduction is interrupted, and the rate of heat flow is reduced. To traverse the air spaces, heat must be transferred by radiation and convection.

2) The width of the air space affects its thermal performance. Up to approximately 5/8 inch, the wider the air space, the greater the reduction in heat flow. An air space narrower than 3/16 inch begins to be ineffective. Across such a short distance, heat is readily conducted by the air. At the other extreme, increasing the air space width beyond approximately 5/8 inch does not substantially reduce the U-value below that of the 5/8 inch separation (although it can substantially improve the acoustic insulation). This is due to a wider space, allowing the air to circulate freely. Air in contact with the warm sheet of glass rises, air in contact with the cold sheet of glass settles, and a cyclic air movement is established. This moving air transports the heat from the warm glass to the cold glass. The increased heat loss due to such convection currents offsets the decreased losses by conduction through the air. The net effect is shown in the following graph of U-values for different air space widths. The upper three curves show the combined heat transfer for convection and conduction for three temperature differences across the air space. (54, Robinson, p11)

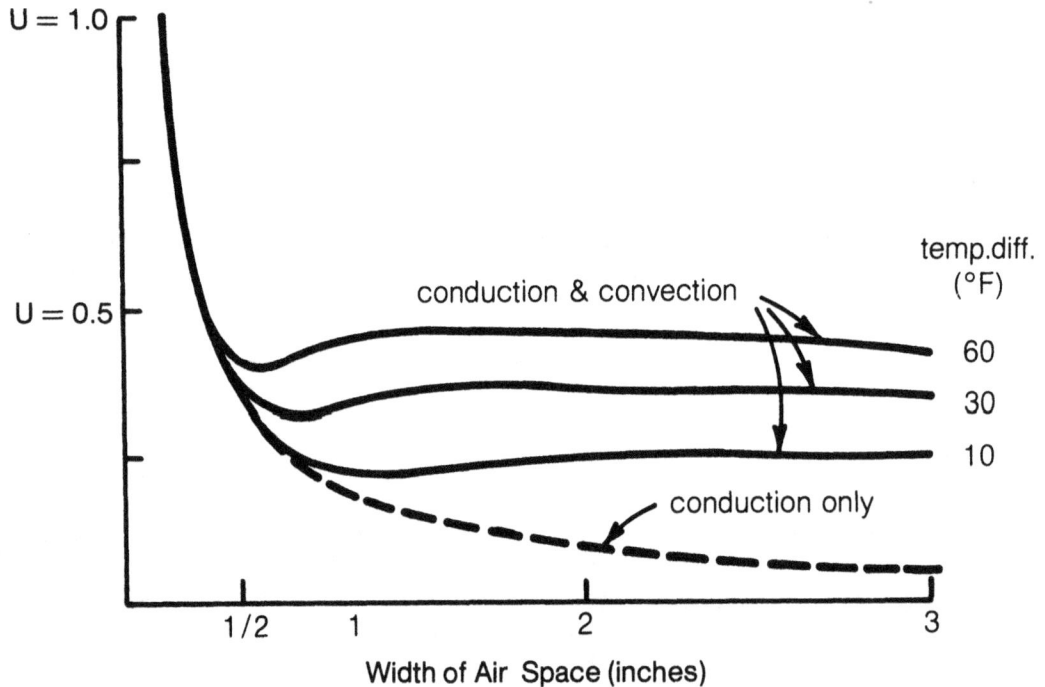

Figure 31.   U-Valved Insulation Glass
vs. Separation

3)   Three layers of glass separated by air spaces are more effec-
     tive than two separated layers of glass of the same overall
     width. Triple glazing with two 1/4-inch air spaces has a
     U-value of 0.47 compared to 0.58 for double glazing with a
     single 1/2-inch air space. (74,ASHRAE,p370) Installing storm
     windows over double-glazed windows is another means of
     achieving triple glazing.

4)   The mix of gases in the air space affects the rate of heat
     transfer. For example, heat flow through an insulating glass
     unit (in a 12-mph wind, an outside temperature of 28° F and an
     indoor temperature of 70° F) will be reduced 14 percent when
     the air space is filled with Krypton. (75,Berman,p33) As
     much as an 18- to 20-percent reduction in U-value may be
     possible with carbon dioxide. (75,PRITSKER)

5)   The heat absorbing and radiating characteristics (charac-
     terized by the termed emittance) of the two surfaces of glass
     facing toward the air space will affect the rate at which heat
     is radiated across the cavity. A coated film, such as tin
     oxide or indium oxide, or pure metals, such as gold, silver,
     or copper, applied to either of the glass surfaces facing the
     cavity reduce the heat transfer by radiation. Examples of the
     effectiveness of such coatings is given in the following table
     (74,ASHRAE,p370)

4-2

```
Type of Insulating Glass                          Winter U-value
(1/2" air space)
effective emittance = 0.8 (untreated). . .   0.58

effective emittance = 0.60 . . . . . . . .   0.52

effective emittance = 0.40 . . . . . . . .   0.45

effective emittance = 0.20 . . . . . . . .   0.38
```

Note that the U-value for glass having the lowest emittance coating is comparable to uncoated triple glazing separated by two 1/2-inch air spaces (U = 0.36). (74,ASHRAE,p370)

Reflective coatings applied to the inside surface of the outer sheet can also reduce the emissivity of the glass surface and, hence, reduce the radiation across the air space. A highly reflective glass can thus reduce the U-value to as low as 0.29. (77,PPG,p15)

The benefit of such coatings in reducing winter heat conduction outward must be balanced against a loss in daylight transmittance and loss of natural solar heating and illumination.

6)   Too great a difference between body temperature and the temperature of nearby surfaces results in a high rate of radiant heat loss and resulting discomfort. A study in England indicates that people in a room feel uncomfortable near surfaces having a temperature 8° C (14.4° F) above or below the average temperature of all other surrounding surfaces. (75,McIntyre,p6) Because the heat flow is substantially reduced, the inside surface temperature of insulating glass is much closer to room temperature than is the case with single glass, and so discomfort near windows is alleviated. This benefit is also realized with storm windows.

7)   In the summer, multiple glazing reduces the amount of heat conducted from the outdoors inward. The U-value for summer conditions is slightly lower for single glass, and higher for insulating glass, due to changes in air-space convection at higher temperatures, because a lower wind speed is assumed. However, multiple glazing reduces the amount of sunlight transmitted. Because glass is not 100-percent transparent, some of the sunlight is absorbed and converted to heat within the glass. This heat is then dissipated to the air at both surfaces of the glass and radiated from both surfaces. The heat which is dissipated to the outdoors is heat with which the air conditioning system never has to contend.

Shading coefficients are a relative measure of the total solar heat transferred to the interior. The performance of single sheet glass (of double strength) is used as the basis of comparison and hence has a shading coefficient of 1.00. The shading coefficient of a window with a storm sash or insulating glass is reduced to approximately 0.90. This can further be reduced by using reflective or tinted glasses in the outer glazing. For example, heat absorbing glass could reduce the shading coefficient to 0.56. (74,ASHRAE,p400) If the outer sheet is reflective glass the shading coefficient is reported to be as low as 0.17. (75,LOF,p19)

ADVANTAGES:

1) Reduced conducted heat loss in winter, both during the night and during the day, when the window can perform as a solar collector.

2) Slightly reduced solar transmission and hence slight reduction in air conditioning load.

3) Improved winter comfort. The inside surface of the glass is closer to room temperature, so it is no longer uncomfortable to be near the window.

4) Elimination of condensate or ice forming on the glass except in extreme weather.

5) Reduced sound transmission. (75,Sabine,p.27)

6) Reduced infiltration possible with storm sash.

DISADVANTAGES:

1) Greater weight compared to single glass windows. The two layers of glass make the window nearly twice as heavy. This makes installation more difficult, and removal of sash for cleaning more awkward.

2) Unsightliness, if the seal becomes leaky condensate will form on the inaccessible glass surfaces within the air space. (Manufacturers guarantees against such leakage for a specified period of time.)

3) Replacement of broken glazing is more costly and time consuming when non-stock sizes necessitate special ordering.

4) No reduction in infiltration by installing insulating glass, as contrasted to installing storm windows, which provide an additional layer to resist infiltration.

5) Increased likelihood of thermal breakage due to the higher glass temperature in the presence of sunlight. (Manufacturers recommendations for installation must be carefully adhered to.)

AESTHETICS:

1) The presence of insulating glass instead of a single pane of glass is nearly indiscernable visually. A storm sash, if the frame is the same color and is congruent with the underlying window, is unobtrusive but does reduce the setback of the glass on the facade.

2) Low emission, heat absorbing, or reflective glass used in the outer sheet of insulating glass changes the light transmission characteristics of the window. Various tints may be specified, ranging from grays to bronzes. The visual and other effects of these glasses are discussed more completely in the section on reflective glasses.

COSTS:

The following is a comparison of glazing costs based upon estimates in the Washington, DC, area, for small quantities without installation.

Single glazing	3/16" thick	$1.00/ft^2
Double glazing	5/8" overall	$4.50/ft^2
Storm sash		
Prime window	3/16" thick	$1.00/ft^2
Storm sash with frame	single strength	$2.00/ft^2
Triple glazing		
Prime window	5/8" overall	$4.50/ft^2
Storm sash with frame	single strength	$\underline{\$2.00/ft^2}$
		$6.50/ft^2

It should be noted that although there is a substantial difference in glazing costs when multiple glazing is specified, the cost of the total window does not increase proportionally. Typically, a total window unit consisting of frame, sash, and glazing increased in price approximately 25 percent for double versus single glazing, and approximately 45 percent for triple versus single glazing.

EXAMPLES:

1) A study by the Edison Electric Institute indicates that a typical, well-insulated, all-electric, ranch style house could save 3,266 kWh of electricity with insulating glass instead of single glass in a climate area such as Indianapolis (5,611 degree days). At an electric rate of $0.04/kWh, this amounts to $130.64 per year. This represents savings for heating costs only. (76,PPG,p4)

2) A computer study of two Fairfax County, Virginia, schools indicates a 13-percent and 10-percent savings for two buildings reglazed with insulating glass. (74,Griffin,p67)

3) A study in Sweden calculated the energy savings from triple glazing compared to double glazing. The study was based on a hypothetical office module with other offices above, to the sides, behind, and below. The glass area was 2.24 m^2 (24.1 ft^2). The office was occupied by two people from 0800 to 1600 hours. Their heat output combined with electrical equipment was assumed to be 300W. Ventilation was supplied at a rate of 80 m^3/hr (2825 ft^3/hr). Room temperature is kept at 22° C (68° F). The following table summarizes the average daily energy saved, triple versus double glass. (75, Adamson,p11)

Orientation	Place	kWh for		Energy Saved		
		double	triple	kWh/yr per window	kWh/yr per m^2	(ft^2)
North	Malmo	3490	3290	200	89	(8.3)
	Stockholm	4140	3910	230	103	(9.6)
	Lulea	5880	5550	330	147	(13.7)
East	Malmo	3110	2940	170	76	(7.1)
	Stockholm	3770	3570	200	89	(8.3)
	Lulea	5400	5120	280	125	(11.6)
South	Malmo	2650	2530	120	54	(5.0)
	Stockholm	3320	3150	170	76	(7.1)
	Lulea	4950	4710	240	107	(9.9)

Note: Malmo: 6,900 degree days.
       Stockholm: 7,700 degree days.
       Lulea: 11,000 degree days.
(77,Helander)

REFERENCES:

Adamson, Bo and Kallblad, Kurt, Time for Triple Glazing,
Department of Building Science, Lund Institute of Technology,
Lund, Sweden, 1975.

American Society of Heating, Refrigerating and Air Conditioning Engineer, ASHRAE Handbook of Fundamentals, ASHRAE,
Inc., NY, 1972.

Anques, J. and Croiset, M., Thermal Comfort Requirements
Adjacent to Cold Walls - Application to Glazed Openings,
Centre Scientifique et Technique du Batiment, Paris, France,
English translation: NBS Technical Note 710-4, NBS, Washington,
DC, 1972.

Berman, Samuel, Energy Conservation and Window Systems.
N.T.I.S., Springfield, VA, 1975.

Griffin, C. W., Energy Conservation in Buildings: Techniques
for Economical Design, Construction Specifications Institute,
Washington, DC, 1974.

Helander, Lars, telephone conversation, Scientific Attache,
Swedish Embassy, Washington, DC, Feb. 15, 1977.

Libbey-Owens-Ford Co., Glass for Construction, LOF Co., Washington, DC, 1975.

McIntyre, D. A., Radiant Draughts, Electricity Council Research Centre, Capenhurst, Chester, England, 1975.

PA, Technics: High-performance Glass, Progressive Architecture, Reinhold Publications, Stamford, CT, June 1976.

PPG Industries, Inc., Architectural Glass Products, PPG Industries, Inc., Pittsburgh, PA, 1977.

PPG Industries, Inc., Energy Effective Windows for Cost Savings and Conservation, PPG Industries, Inc., Pittsburgh, PA, 1976.

Robinson, H. E. and Powelitch, F. J., "The Thermal Insulation Value of Air Spaces", Housing Research Paper No. 32, Housing and Home Finance Agency, Washington, DC, April 1954.

Pritsker, Theodore, Assoc. Dallas Laboratories, Dallas, Texas Telephone conversation, Dec. 3, 1975.

Robinson, Henry E., Editor, Calculation of Pane-to-Pane Temperature Difference, Durability of Insulating Glass, BSS 20, National Bureau of Standards, Washington, DC, 1970.

Sabine, Hale and Myron, Lacker. Acoustical & Thermal Performance of Exterior Residential Walls, Doors, and Windows, NBS BSS 77, 1975.

Shand, E. B., Glass Engineering Handbook, McGraw Hill Book Co., NY, 1958, p.27.

Stephenson, D. G., Heat Transfer at Building Surfaces, Canadian Building Digest, National Research Council of Canada Ottawa, April 1964.

4.2 HEAT-ABSORBING GLASS/Shading, Solar Heating

STRATEGY:

Install glass which absorbs more solar energy than clear glass to reduce solar heat gain.

PHENOMENA:

1) Visible light represents only a part of the total solar radiation. Solar energy at sea level is comprised of approximately 3 percent ultraviolet, 44 percent visible, and 53 percent infrared energy. (74,ASHRAE,p387) All of this energy, when absorbed, is converted to heat. Therefore, it is the amount of total solar energy transmitted which determines the amount of heat gain, and the amount of visible light transmitted which determines the amount of illumination provided.

2) Adding a metallic oxide to the ingredients of glass during its manufacture increases its absorptivity of visible and near infrared solar energy. The greater absorptivity occurs in the near infrared range. This characteristic distinguishes heat absorbing glass from glass which is merely tinted. (73,LATTA,p61) This is an advantage in that the visible light from the sun provides illumination which must otherwise be provided by more heat intensive electric lighting.

3) The solar energy absorbed by the glass becomes heat which is radiated and convected to the outdoors and indoors proportional to the temperatures, air movements, and the surface characteristics of either side of the glass,(if different). Unfortunately, on a still, sunny, summer day more heat is dissipated indoors because the air conditioned building interior is cooler. Conversely, in the winter more heat is dissipated to the outdoors because the outside temperatures are lower than the inside temperature. The following figure shows that heat-absorbing glass is an improvement over single glass but still admits much of the summer suns heat. The percentages given in the figure are for an example case and will vary as the sun angle varies. (64,Ulrey,p168)

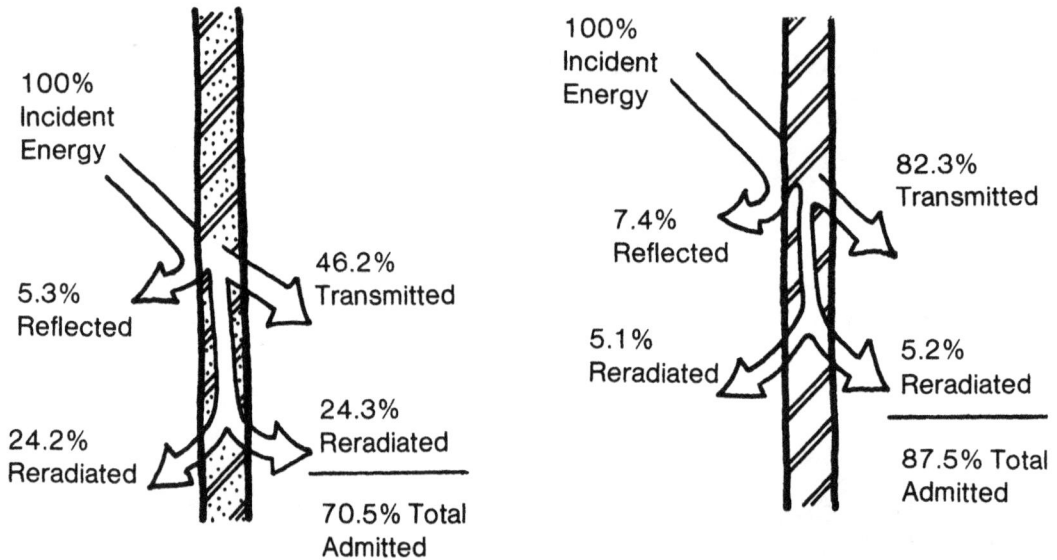

Figure 32. Solar Energy Transmission Through Heat-Absorbing
Single Glazing vs. Clear Glass (SUMMER)

3) When heat absorbing glass is used as the outer sheet of
    glass in double glazing, its performance is substantially
    improved compared to its use as single glazing. In order
    for the heat in the glass to enter the building it must
    first bridge the trapped air space by radiation and
    convection, be conducted through the inner sheet of
    glass, and then be radiated and convected into the build-
    ing interior. More heat will be dissipated to the out-
    side air which is in direct contact with the heat absorbing
    glass. Furthermore, the outward rate of heat dissipation
    greatly accelerates if there is any wind. (73,LATTA,p60)

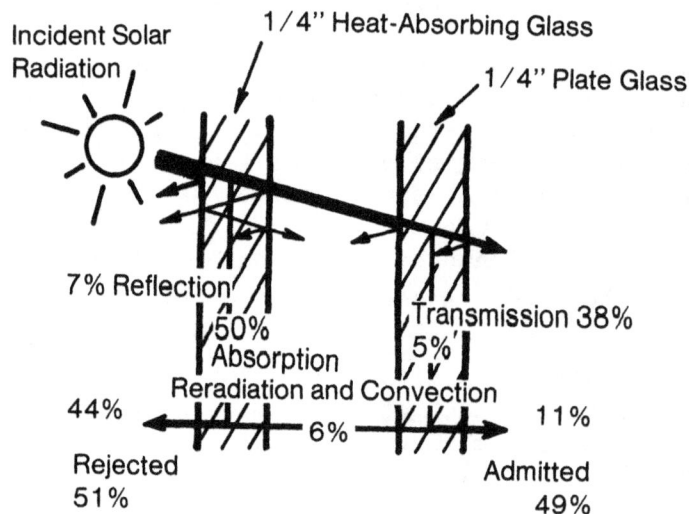

Figure 33. Solar Energy Transmission
Heat Absorbing Insulating Glass

4) The following statistics are examples of the performance of heat-absorbing glass in single and double glazing configurations as reported by one glass manufacturer. (75,LOF,p19) Exact values will vary depending on the composition of the glass.

PERFORMANCE OF HEAT-ABSORBING GLASS

GLASS	VISIBLE TRANSM	TOTAL SOLAR TRANSM	SHAD. COEFF.
1/4" CLEAR	88%	77%	0.93
1/4" HEAT ABS.	75	47	0.70
1" CLEAR INSUL.	77	59	0.79
1" HEAT ABS. INSUL.	66	36	0.56

5) If heat-absorbing, double glass is installed in a reversible sash, the heat absorbing sheet of glass can face the outside in summer to dissipate heat outward, then be reversed in the winter so that the heat-absorbing glass faces the inside dissipating its heat into the building. The effectiveness of this configuration could be even further increased by providing closable vents above and below the heat-absorbing glass. In the winter, during the hours of sunlight, the vents could be opened to circulate the heated air between the two sheets of glass into the building interior, then closed at night to preserve the insulating value of the air space. In the summer with the sash reversed, the heated air between the glass sheets could be discharged to the outdoors further reducing the amount of heat which enters the building. The following diagram illustrates the effects of seasonally reversed sash.

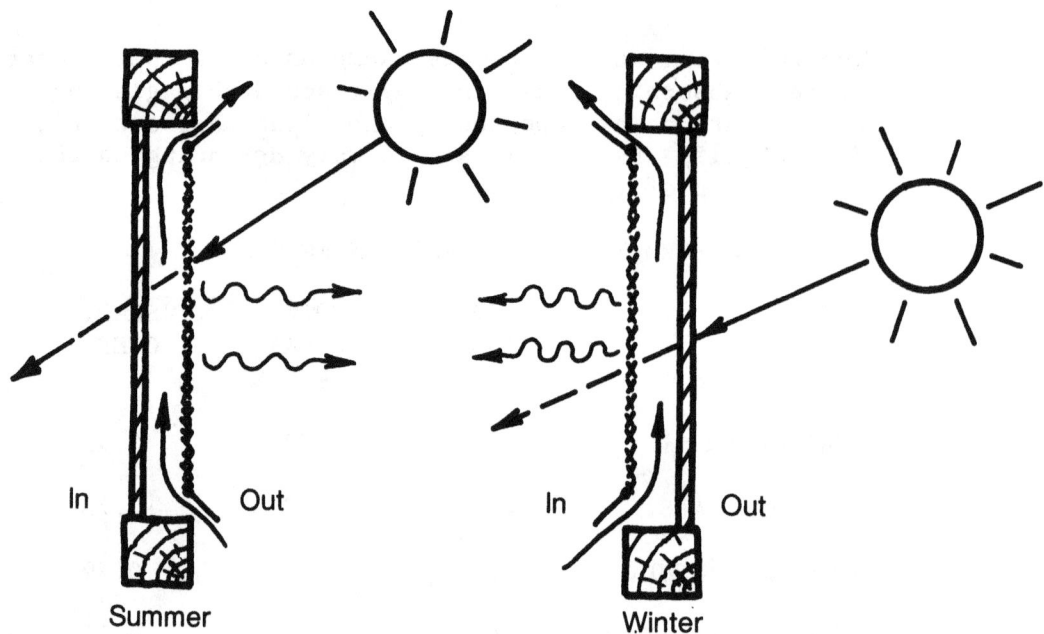

Figure 34. Reversible Sash With Heat-
Absorbing Double Glass

6) Since sunlight absorbed in heat-absorbing glass almost
immediately raises its temperature, the time lapse
between window exposure to sunlight and room temperature
rise is very short. Alternatively, with clear glazing
most of the sunlight is transmitted through the window
and absorbed by the walls, floor, and furniture within
the room. Thus the room temperature rise is delayed
while heat from the transmitted sun is absorbed into the
mass of these objects. (See STRATEGY: Thermal Mass)

7) Closed draperies, roll shades, or blinds can reflect
much of the sunlight which has penetrated heat-absorbing
glass back at the glass. This double exposure of the
heat-absorbing glass to sunlight, substantially increases
the glass temperature. Similarly, when heat absorbing
glass is used in double glazing, the inner sheet of
clear glass reflects part of the sunlight back to the
outer sheet of heat-absorbing glass increasing its
temperature. These high temperatures create large
stresses within the glass.

ADVANTAGES:

1) Reduced summer solar heat gain largely by absorption of
non-visible solar radiation and to a lesser extent by
absorption of visible light which provides illumination
and view.

4-12

2) Reduced fading of fabrics due to greater absorption of ultraviolet solar radiation compared to clear glass.

3) Winter solar collection, summer solar rejection with reversible, double glazed sash with clear and heat-absorbing glass and provision for operable venting at the top and bottom of the sash.

DISADVANTAGES:

1) Partial dissipation of heat to the indoors in the summer and to the outside in winter when heat-absorbing glass used in single glazing.

2) Possible breakage of heat-absorbing glass when drapes or shades drawn in the summer. The strength of the glass at the edges is especially critical in such instances.

3) Increased temperature of glass increases radiation of heat which increases likelihood of discomfort for occupants near windows.

AESTHETICS:

1) Which of the many available metallic additives are added to the ingredients of the glass during its manufacture determines the tint it will have. For example, iron oxide imparts a bluish green color. Nickel and cobalt oxides and selenium give a gray or bronze tint. (75, Architects Journal,p1263) The designer must consider both how this tint will alter the colors of the building interior and, how the glass color will harmonize with the other colors of the building exterior.

2) The view out through heat-absorbing glass is dimmer than through clear glass but brighter than many of the reflective glasses.

3) The thickness of heat-absorbing glass will affect its color since the tinting is caused by an ingredient dispersed throughout the glass rather than occurring only at the surface as with reflective glass. This means that if various window sizes dictate different glass thicknesses for reasons of strength, the color-density variation as viewed from the outside, and the brightness variation as viewed from the inside, must be considered in the architectural composition, otherwise the thicker size should be used throughout.

COSTS:

Heat-absorbing glass costs approximately 1/3 to 1/2 more than clear glass.

EXAMPLE:

The effectiveness of heat-absorbing glass, though superior to clear glass in many instances, is generally inferior to reflective glass for the purpose of shading solar radiation. The following study done in Switzerland compares the performance of single clear, single heat-absorbing, and double reflective glazing with plastic draperies. Six south-west facing rooms comparable in every respect were used in the study. Each room had 146 sq. ft. of floor area with a 9.25 ft. ceiling height, 102 sq. ft. of outside wall area, and 46 sq. ft. of window area divided between two windows (approximately 50 percent window to outside wall ratio). The outside air temperature ranged from 20 to 23°C (68-74°F). Average room air temperatures and inside glass surface temperature were recorded for September 10, 11, and 12 of 1969. (74, Granjean, p206)

Glass	Room air temp.	Glass temp.
Clear	29.9° – 35.4°C (86° – 96°F)	35.8° – 39.6°C (97° – 103°F)
Heat Absorbing	28.7° – 32°C (84° – 90°F)	38.3° – 46.2°C (101° – 115°F)
Double Reflective	23° – 27°C (74° – 81°F)	24° – 34°C (75° – 94°F)

The performance of the heat-absorbing glass would be better if it were double glazing as was the case with reflective glazing. The room temperatures would be less extreme in all cases if the building had more thermal mass (the building studied was light-weight construction).

REFERENCES:

A. J., "Window Glass Design Guide", The Architects Journal, Architectural Press Ltd., London, Jan. 1976.

Dubin, Fred, "Energy for Architects", Architecture Plus, Informat Publishing Corp., New York, N. Y., July 1973.

Grandjean, E., The Ergonomics of the Home, Halsted Pr., N. Y. 1974.

Latta, J. K., Walls, Windows and Roofs for the Canadian Climate, Division of Building Research, National Research Council Canada, Ottawa, Canada, 1973.

LOF, "Glass for Construction", Libbey-Owens-Ford Co., Toledo, Ohio, 1975.

Persun, Charles I., "Gray Glass", Progressive Architecture, Reinhold Publishing Company, Stamford, Conn., April 1958.

Stephenson, D. G., and Mitalas, G. P., "An Analog Evaluation of Methods for Controlling Solar Heat Gain through Windows" Research Paper No. 154, Division of Building Research, National Research Council Canada, April 1962.

Ulrey, Harry F., Audels Architects and Builders Guide, Theodore Audel and Co. Indianapolis, Ind., 1964.

4.3 REFLECTIVE GLASS/Shading

STRATEGY:

Install glass with a reflective surface to reduce summer
solar heat gain.

PHENOMENA:

1) Solar energy striking a window is either reflected,
    absorbed, or transmitted. By increasing the amount of
    solar energy reflected, the amounts absorbed and trans-
    mitted are reduced. Solar energy absorbed in the glass,
    or transmitted and absorbed within the building, becomes
    heat. Reflective glass, by increasing the amount of
    solar energy reflected at the window, therefore, reduces
    the eventual air conditioning load within the building.

2) Reflective glass used as the outer sheet of insulating
    glass is more effective at keeping out the sun's heat
    than reflective glass used as single glazing. This is
    due to the fact that reflective glass absorbs more sun-
    light than clear glass and its temperature rises. When
    this heat is concentrated in the outer sheet of insu-
    lating glass it is more easily dissipated to the outside
    air, especially if there is a breeze. Additionally, the
    trapped air space acts as insulation impeding the inward
    flow of heat. The following table illustrates the
    effectiveness of reflective glass used as single glazing
    and as the outer sheet of insulating glass compared to
    the performance of clear single and double glass.
    (75,LOF,p19)

TYPE OF GLASS	VISIBLE TRANSM.	TOTAL SOLAR TRANSM.	SHAD. COEFF.
1/4" clear single	88%	77%	0.93
1/4" gray reflective single	34	36	0.60
1" clear insul.	77	59	0.79
1" gray reflec. insul.	30	29	0.47

3) The importance of the reflective coating or film occuring on the outer sheet of insulating glass is illustrated in the following figure. (73,LATTA,p60)

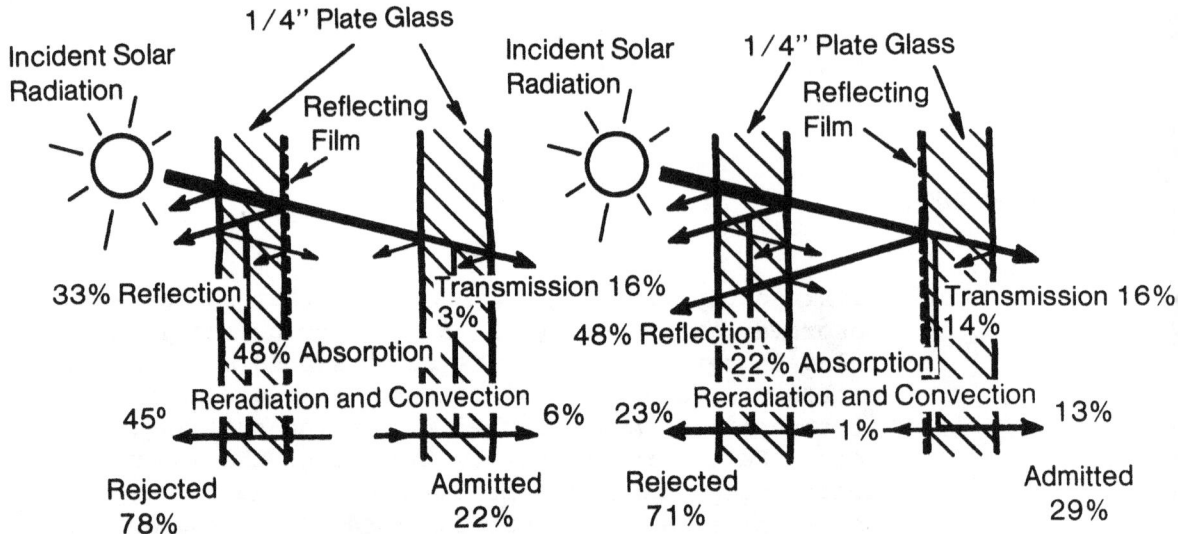

Figure 35. Effect of Film Location
on Heat-Gain

4) The use of certain types of reflective coatings on one of the sheets of insulating glass can reduce the radiation of heat across the air space and hence reduce winter heat losses. The winter U-value of clear insulating glass with a 1/2 inch air space is 0.58 versus as low as 0.28 for insulating glass with the outer sheet of reflective glass. (76,PPG,p15) (See STRATEGY: Applied Films.)

ADVANTAGES:

1) Reduced summer solar heat load.

2) Reduced likelihood of glare.

3) Reduced heat loss in the winter for certain types of reflective glass.

DISADVANTAGES:

1) Decreased transmitted solar energy in the winter and decreased daylight illumination year-round.

2) Care required to avoid scratching reflective coating when washing windows. Reflective single glazing frequently has the reflective coating on the inside surface of the glass.

3) Caution required in selecting a sealant to avoid problems due to lack of bonding with reflective coating of the glass. (Single glazing only.)

4) Increased replacement cost in the event of breakage by vandals, storm damage, or other causes. (Also true for insulating glass and heat-absorbing glass.)

5) Caution required in detailing the setting of reflective glass to avoid thermal breakage. The higher solar absorptivity of reflective glass compared to clear glass results in higher surface temperatures, especially with insulating glass. If the glass is installed in a massive material such as concrete, the slowness of the concrete to heat up in comparison to the glass when sunlit will result in extreme center to edge temperature differences and edge stresses in the glass. One means of reducing these stresses is to seat the glass in a rubber gasket. The rubber serves as an insulation which reduces the conduction losses at the edge of the glass and allows the edges to be closer to the temperature of the center areas of glass. It is also extremely important that if glass must be cut in the field, special provisions be taken to insure a clean cut edge. (See Manufacturer's literature for instructions on cutting glass)

AESTHETICS:

1) Because an outside viewer looks at reflective glass rather than through it distortion is more critical than is the case with clear glass. Trees and clouds, because of their soft geometry is less problematic than reflection of adjacent buildings. Full scale mock-ups on the site are worthwhile for studying the visual appearance of reflective glass and whether distortion is bothersome or not. (77,Skolnik,p93)

2) A reflective glass building may create glare for the occupants of adjacent buildings as well as pedestrians and drivers.

3) Reflective glazing of isolated individual windows may appear awkward. Reflective glazing tends to be more visually effective in expanses, either as a series of windows in continuous band, or a single large area.

4) Where the fenestration carries around a corner or occurs on opposite sides of narrow parts of a building, such as elevator lobbies or bridging corridors, building the transparency achieved with clear glass is not possible with reflecting glass.

5) Reflective glass darkens the view out by virtue of the reduction in light transmission. Colors may also tint the view as in the case of bronze or other tinted reflective glasses. This darkening of the view and slight tinting are not likely to be distracting unless a window with clear glass is in the vicinity, or unless the coated surface of the reflective glass is scratched.

COSTS:

EXAMPLE:

In the design of the Toledo Edison Building the architects, along with the glass manufacturer, conducted a detailed computer study of the effects of a variety of glasses on the buildings' construction and operating costs. They selected a chromium-coated, dual-wall insulating glass which increased the first cost of the glass by $122,000 compared to conventional 1/4" float glass (float glass has largely replaced plate glass). However, offsetting this first cost of the glass was a savings of $123,000 in initial costs for the heating and cooling equipment and ductwork. (A 64.7 percent reduction in the capacity of the central refrigeration system, a 53.2 percent reduction in the capacity of the central heating equipment and a 67.9 percent reduction in the capacity of the distribution system.) Resulting energy consumption savings were calculated to be 729.4 kilowatts per hour which translates to a savings in yearly operating cost of approximately $40,000. (73,NBS,p84)

REFERENCES:

American Society of Heating, Refrigerating & Air Conditioning Engineers, ASHRAE Handbook of Fundamentals, ASHRAE Inc., N. Y., 1974.

C-E Glass, "C-E Glass", Combustion Engineering, Inc., Pennsauken,
    N. J., 1975.

Latta, J. K. Walls, Windows and Roofs for the Canadian Climate,
    Division of Building Research, Ottawa, Canada, 1973.

LOF, "Glass for Construction", Libbey-Owens Ford, Toledo, Ohio,
    1975.

Kusuda, T. and Ishii, K. Hourly Solar Reduction Data for the
    Average Days National Bureau of Standards, Wash., D. C., 1976.

NBS, "Technical Options for Energy Conservation Buildings." NBS
    Tech. Note 789 National Bureau of Standards, Wash., D. C., 1973.

PA, "Technics: High-performance Glass", Progressive Architecture,
    Reinhold Publications, Stamford, Conn., June 1976.

PPG Industries, Inc., "Architectural Glass Projects", PPG Industries
    Inc., Pittsburgh, Pa., 1976.

Skolnik, Alvin D., "Aesthetic Evaluation of Glass", Progressive
    Architecture, Reinhold Publishing Co., Stamford, Conn., Jan.
    1977.

STRATEGY:

Apply a reflective or low-emissivity film to the inside surface of glass to reflect sunlight back out a window but let the view in.

PHENOMENA:

1)   Metalic oxides deposited on transparent plastic films are available which reflect much of the incoming solar energy while still permitting a view out. The shading coefficient of such reflective films can be as low as approximately 0.24.

2)   Other types of film coatings are available which increase the window's reflectivity of room temperature infrared heat while only minimally reducing the window's transparency to the beneficial incoming solar energy. The net effect of such "low-emissivity" films is a reduction in the winter U-value of the window from 1.13 to as low as 0.74. (75, Berman,p61)

3)   Solar reflective films and to a lesser extent low-emissivity films have the disadvantage of reducing the beneficial aspect of incoming solar energy in the winter. The following table illustrates the seasonal benefit (+) or detriment (-) of single glazing compared to glazing with applied low-emissivity or reflective films in a southern and northern city. (75,Berman,p64)

WINDOW ENERGY DEMAND BY ORIENTATION (KBTU/SQ. FT.)

CITY	NORTH		EAST		SOUTH		WEST	
	Wntr	Sumr	Wntr	Sumr	Wntr	Sumr	Wntr	Sumr
Dallas								
No Film	-24	-95	+26	-156	+102	-118	+26	-185
Low-emissivity	-7	-80	+35	-135	+107	-101	+38	-161
Reflective	-46	-41	-33	-56	-14	-46	-33	-63
New York								
No Film	-84	-43	-38	-76	+29	-59	-38	-81
Low-emissivity	-43	-39	-2	-68	+58	-53	-2	-73
Reflective	-105	-11	-93	-19	-76	-15	-93	-20

4) From the table in the previous paragraph it is evident that the annual energy effectiveness of low-emissivity or reflective films will be very dependent on whether air conditioning or heating is dominant, and which direction a window faces.

Due to the dominance of air conditioning in Dallas reflective films conserve energy on north, east, and west-facing windows while in New York may be are only minimally conserving for these orientations. In both Dallas and New York reflective films would appear to result in increased energy consumption when applied to south-facing windows.

Low-emissivity films appear to be more energy conserving than no film or reflective film only on south-facing windows in Dallas but on all exposures in New York.

ADVANTAGES:

1) Reduced summer solar heat gain with reflective films.

2) Reduced glare without obscuring view.

3) Can be applied to existing windows.

4) Reduced fabric fading.

5) Reduced winter heat loss with "low-emissivity" films.

6) May hold glass together in event of shattering (if film thickness adequate).

DISADVANTAGES:

1) Reduced benefit of daylight and solar heat gain in winter with reflective and to lesser extent selective films.

2) Replacement of film is costly, difficult, and likely to be required after nine to twelve years.

3) Additional care required in washing to avoid scratching surface of films which are softer than glass.

4) Reduced effectiveness of selective coatings when applied to single glazing because the need for abrasion resistance necessitates a protective coating over the selective coating.

5) Possible cause of breakage when installed as a retrofit item. (See disadvantages listed under Strategy: Reflective Glass)

AESTHETICS:

1) Reflective films require the same aesthetic judgment as discussed in the strategy: "Reflective Glass".

2) Selective films may impart a slight tint to the outside view.

COSTS:

1) Reflective films cost approximately $0.40 to 0.50 per square foot depending on the properties and quantity specified. Installation ranges from $0.70 to 0.90 on the same basis. (75,GSC,p9)

2) Low-emissivity films have not been used as widely in window applications but the product is commercially available. A polyester film with a gold coating is one potentially effective product currently marketed by at least two manufacturers for other uses. The price is approximately $1.20/square foot. (76,Sierracin,p1) (76,Levy,telephone)

EXAMPLES:

1) A reflective film applied to windows in an office building in Silver Spring, Maryland resulted in a 50 percent reduction in the air conditioning load. The average air conditioner operation cycle was reduced for 24 hours per day during peak summer periods to 12 hours per day. Winter heat loss reductions were also observed but not quantified. (75,Groves)

2) A five mil, polyester film with a coating of gold is currently available which transmits up to 80 percent of the visible light but reflects over 95 percent of room temperature heat back into the room. The material is currently being marketed for space suit visers and ski goggles. (76,Sierracin)

REFERENCES:

Berman, Samuel M., "Energy Conservation and Window Systems", PB-243-117, National Technical Information Service, Springfield, Va., Jan. 1975.

Cellarosi, Mario, telephone conversations, Physical Properties of Glass Section, National Bureau of Standards, Washington, D. C., Dec. 10, 1976.

F.E.A., UCAN Manual of Conservation Measures, Conservation Paper Nr. 35, Federal Energy Administration, Washington, D. C., Nov. 1975.

GSC, "Solar Master Controls Heat, Glare & Fade", General Solar Corp., Rockville, Md., 1975.

Groves, Roy L., Letter to General Solar Corp. Rockville, Md., L. L. Rust Co., Washington, D. C., May 1, 1975.

Levy, Donald, telephone conversation about "Lockspray-Gold", Lockheed Corp., Palo Alto, Cal., Dec. 13, 1976.

Levy, D. J., "The Lockspray-Gold Process, A technical Bulletin", Lockheed Palo Alto Research Laboratory, Palo Alto, Cal., April, 1966.

Levy, D. J. and Momyer, W. R., "Spray Deposition of Low Emittance Gold Thermal Control Coatings", SAMPE Journal, Society for Advancement of Material & Process Engineering Inc., Azusa, Cal., Apr/May 1969.

Sierracin, "Sierracin Intrex, Electrically Conductive Film Components", Sierracin Corp., Sylmar, Cal., 1976.

4.5   REDUCED GLAZING/Insulation, Shading

STRATEGY:

Reduce the glass area in the wall opening by substituting an insulating panel in a portion of the window.

PHENOMENA:

1)   Single glazing has a U-value of 1.13, double glazing approximately 0.58, and opaque, insulated panels as low as 0.10. (76,PPG,p18) By replacing part of the glass with an opaque, insulated panel the outward flow of heat in the winter is reduced. Also, air conditioning costs may be lowered due to the reduced admittance of sunlight.

2)   The benefit of reduced outward heat flow must be compared against the loss of beneficial winter solar heat and beneficial daylighting. Opaque, insulated panels might be beneficial on north-facing windows but detrimental on south-facing windows. Climate and orientation are important factors.

3)   Opaque, insulated panels are commonly installed in the upper section(s) of windows. The high portion of a window provides the deepest penetration of daylight into a room. Thus, the panels not only reduce the amount of glass area admitting daylight but also reduce the depth the daylight penetrates into a room. Heat loss reductions must, consequently, be considered against the possible additional cost of continuous as opposed to periodic use of electric lighting.

ADVANTAGES:

1)   Reduced heat loss in winter, sunlight transmission in summer.

2)   Reduced window area to wash and maintain.

DISADVANTAGES:

1)   Reduced winter solar heat gain.

2)   Reduced amount and penetration of daylight.

3)   Possible loss of means of egress during fire.

AESTHETICS:

1)   Reduced view of outside. The sky portion of the view is
     cut when panels are installed in upper window sections.

2)   Introduces another material to the facade.

3)   Changes the perceived proportion of windows.

COSTS:

An insulated, opaque panel consisting of a sheet of heat
strengthened glass with a ceramic color fused to the rear
surface, and backed with one-inch fiber glass insulation and a
foil vapor barrier cost approximately the same as clear double
glazing. Aluminum or enameled steel clad panels cost approximately 30 percent less than clear insulating glass. (76,
Hiltman)

EXAMPLE:

A school in Big Fork, Minnesota was expanded from 47,000 to
63,370 square feet and new windows were installed in both the
new wing and throughout the original building. The new window
units consisted of a reduced glass area glazed with insulating
glass and an insulated porcelain panel in the top portion of
the window. As a result of the installation of the new windows,
the total cost to heat the 63,370 square feet with new windows
is lower than the cost of heating the original 47,000 square
feet of school with old windows. (74,Sandstrom,p58) The
following photograph shows the new window system with insulating,
opaque panels in the upper portion of the window. (76,Devac,p4)

Devac, Inc., 10130 State Highway 55, Minneapolis, MN 55441

Further analysis is required to separate the savings realized by the insulated porcelain panels from the savings accrued from the more air-tight frames and double glazing. However, the frequency of outside temperatures of minus 20°F and below in Big Fork, Minnesota suggest that the conduction losses of the glass are not offset by solar heat gains and therefore, the insulating panels used to reduce glass area contribute to the lower operating costs of this particular school. In less severe climates orientation is likely to be a critical determinant of energy savings or even losses realized from replacing glass with opaque insulating panels.

REFERENCES:

Devac, "Devac Windows Cut This Buildings Fuel Bill 34 Percent", Devac Window Co., Minneapolis, Minn.

Hiltman, Michael., telephone conversation, Stolle Corp., Sydney, Ohio, Dec. 9, 1976.

PPG, "Architectural Glass Products", PPG Industries, Inc., Pittsburgh, Pa., January, 1976.

Sandstrom, Harvey E., "Replacement Windows Help Reduce Fuel Bills 34 Percent", American School & University, North American Publishing Co., Philadelphia, Pa., February, 1974.

4.6 GLASS BLOCK/Insulation, Solar Heating, Daylighting

STRATEGY:

Install glass block to admit sunlight and daylight with minimal building heat loss.

PHENOMENA:

1) Glass blocks are hollow, masonry units molded in two halves which are then fused together at a high temperature. When cooled, the permanently sealed air becomes exceptionally dry preventing condensate from forming within the cavity. This air space results in glass blocks having a low U-value, yet admitting approximately 50 to 60 percent of the incident solar energy including 78 to 84 percent of the visible light. (64,Ulrey,p175)

2) The larger the block face dimensions the lower the U-value is. This is due to the fact that the larger size requires the use of fewer blocks for a given area. Heat loss is greatest at the edges of a block because the glass bridges the air space. The following U-values show how larger glass blocks provide better insulation: (75,PC,p10)

NOMINAL SIZE	U-VALUE (single cavity)	U-VALUE (double cavity)
4 x 12 inch	0.60	0.52
6 inch sq.	0.60	--
8 inch sq.	0.56	0.48
12 inch sq.	0.52	0.44

Note that glass blocks are available with double cavities with the same overall block depth - usually a nominal four inches.

3) The greater mass of glass block compared to window glass results in a lag in time between when the sun first falls on the block and when the room temperature rises. The heat gain for west-facing glass block windows is conse-

quently delayed in the morning and conversely, the heat
gain of east-facing glass block windows will be elevated
in the afternoon. To approximate this lag the solar heat
gain factor from the previous rather than the current
hour can be used in calculating heat gain. This product
plus the heat gain or loss through the block (U-value
multiplied by the inside/outside temperature difference)
equals the net heat gain or loss. (74,ASHRAE,p487)

4) The shading coefficient of glass block can be lowered by
contouring the glass surface(s) and/or by fusing various
types of inserts between the two halves before they are
joined in the manufacturing process.

The following table illustrates the effectiveness of
several means of reducting the shading coefficients for
nominal 8 x 8 inch glass blocks. Multiply the coefficients by 1.15 for 12 x 12 inch blocks or by 0.85 for 6 x
6 inch glass blocks. (74,ASHRAE,p408)

TYPE	SHADING COEFFICIENT		
	EXPOSED TO SUN	SHADED N, NW, W, SW	SHADED NE, E, SE
(Window glass)	1.00	--	--
Clear block	0.65	0.40	0.60
Clear with glass fiber insert	0.44	0.34	0.51
Contoured outer surfaces Prismatic inside surfaces and glass fiber insert	0.33	0.27	0.41
Same as above plus Ceramic coating on insert or gray glass used for block or prismatic glass fiber insert.	0.25	0.18	0.27

Note: Shading glass block windows from direct sunlight
substantially reduces heat gain. By providing roof
overhangs calculated to obstruct the summer sun and admit
the winter sun, glass block can be used to the best
thermal advantage.

5) Glare from glass blocks can be reduced with various types of inserts. White opal glass is an example of a glare reducing insert which has the additional benefit of making the block a more uniform brightness. Glare can also be reduced by pressing a grid surface into the glass to diffuse the light.

6) Penetration of light into a room can be increased by casting the inside surfaces of the glass blocks in a prism configuration to direct the light up onto a reflective ceiling. (64,Ulrey,p177)

7) Operating glass units can be composed into glass block panels to provide ventilation.

ADVANTAGES:

1) Solar heat gain and daylighting with greatly reduced heat loss compared to single glazed windows.

2) Control of direction of incoming light with potential to increase penetration into rooms with prism surface blocks

3) Reduced glare with use of inserts or diffusing surface treatment of block.

4) Privacy with diffusing or glare reducing glass blocks.

5) Reduced sound transmission ranging from 35.3 db at 128 cycles per second to 47.5 db at 2048 cycles per second. (64,Ulrey,p175)

6) Vandal resistant. Projectiles which would shatter window glass are deflected by glass block.

7) Forced entry greatly impeded.

8) Fire rated at up to 1 1/2 hours depending upon block type. (75,PC,p8)

DISADVANTAGES:

1) Possible Summer or even Spring and Fall overheating of rooms with large expanses of unshaded glass block and inadequate provision for natural ventilation.

2) Distortion or elimination of view out with most glass block types.

3) Elimination of security surveillance of building interior through glass block windows.

AESTHETICS:

1) Glass block are available in clear units affording only slight distortion to view, and in a variety of patterns which can be combined and arranged in an almost endless number of compositions.

2) Diffusing blocks effectively increase the ambient light level of a room.

3) The joints of glass block impart a grid effect to the fenestration. The scale of the grid can be adjusted by the size and proportion of the block.

COSTS:

The following are a sample of the price of delivered clear glass block:

SIZE	UNIT COST	SQ. FT. COST
6 x 6	$1.75	$7.00
8 x 8	2.25	5.06
12 x 12	4.50	4.50

EXAMPLES:

1) The following two examples illustrate the light penetration with glass block on a south exposure compared to glass block on both a south and north exposures 90 feet apart. Both cases are for 40 degrees N. latitude at 10 a.m. and 2 p.m. on March 21. (66,IES,p7-11)

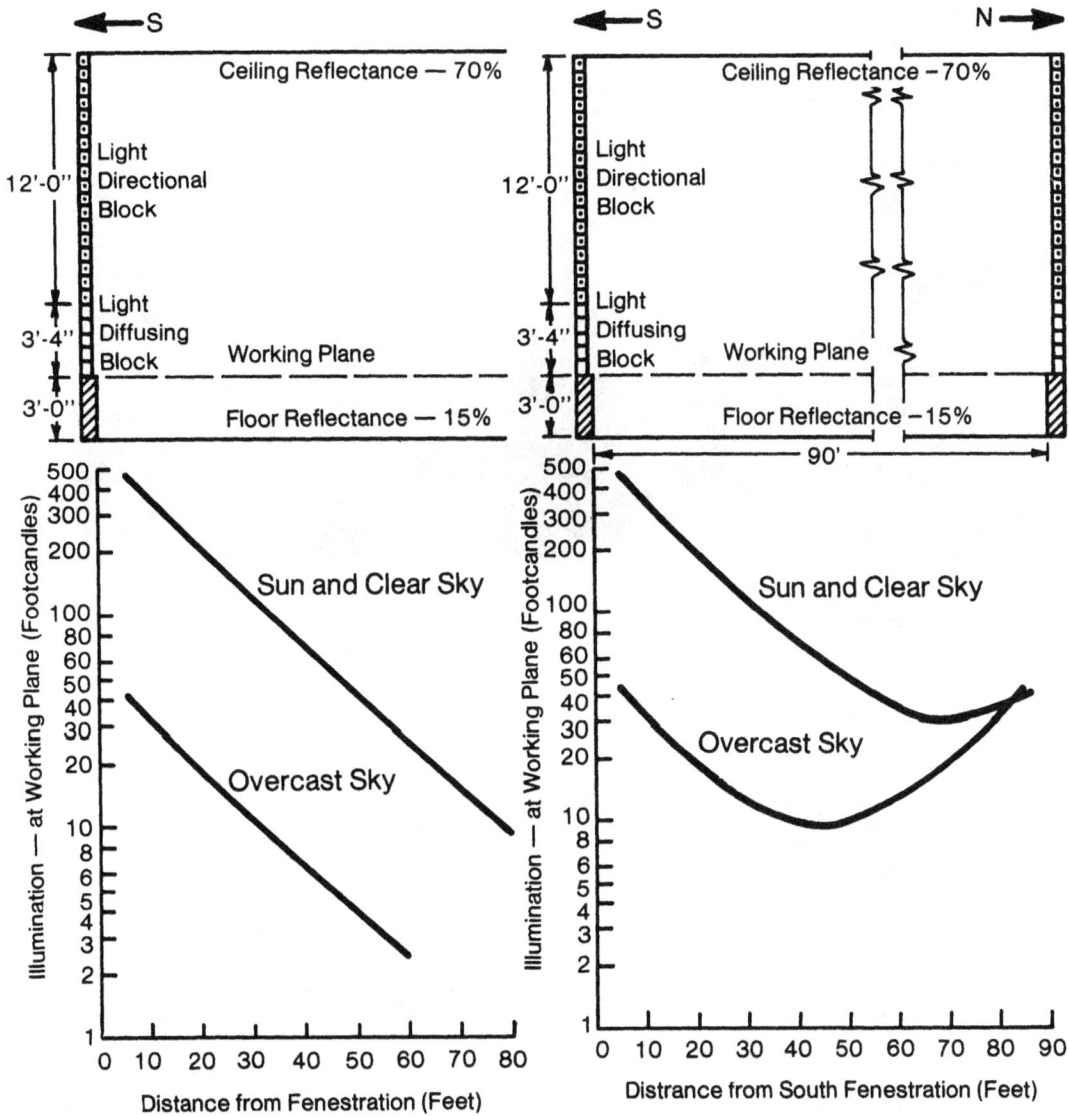

Figure 36. Glass Block on One Exposure vs. Two Exposures

2) Glass block is effective on north exposures to increase the ambient light level of a room while providing a fairly low U-value for heat loss. The following illustration shows the use of glass block to increase the light level of a room and reduce glare from clear glass windows. (75,PC,p15)

Gwathmey Siegel, 154 W 57th St., N.Y. 10019

REFERENCES:

ASHRAE, ASHRAE Handbook of Fundamentals, American Society of
    Heating, Refrigeration, and Air Conditioning Engineers, Inc.,
    New York, 1974.

Boyd, Robert Allen, "The Development of Prismatic Glass Block and
    the Daylighting Laboratory", Engineering Research Bulletin Nr.
    32, University of Michigan, Ann Arbor, Mich., 1951.

Holton, John K., "Daylighting of Buildings - A Compendium and Study
    of Its Introduction and Control." NBSIR 76-1098, National
    Bureau of Standards, Washington, D. C., 1976.

IES, IES Lighting Handbook, Illuminating Engineers Society, New
    York, 1966.

P.C., "Glass Block - Decorative and Functional Units for New
    Construction and Remodeling", Pittsburgh Corning Corp., Pitts
    burgh, Pa., Dec. 1975.

Smith, W. A., and Pennington, C. W. "Shading Coefficients for Glass
    Block Panels", ASHRAE Journal, ASHRAE Inc., New York, N. Y.,
    Dec. 1964.

Ulrey, Harry F., Architect's and Builder's Guide, T. Audel Co.,
    Indianapolis, Ind., 1964.

4.7 THRU-GLASS VENTILATORS/Ventilation

STRATEGY:

Install a transparent wind-driven rotor with a closable louver in a hole cut in a fixed glass window to provide controlled ventilation.

PHENOMENA:

The transparent rotor spins admitting or exhausting air into or out of a room as a result of air pressure differences between the inside and outside. Operable louvers allow the air flow to be stopped entirely when ventilation is not desired.

ADVANTAGES:

1)   Ventilation possible with fixed glass windows.

2)   Reduction of strong sudden drafts during gusty winds.

3)   Security. A six or eight inch diameter hole in a fixed glass window limits the size of burglars who can gain entrance.

DISADVANTAGES:

1)   Minimal natural ventilation compared to operable windows or even compared to an equal diameter unobstructed hole cut in the glass.

2)   Cannot be installed in factory sealed double or tripple glazed windows, nor in existing tempered glass windows by the homeowner or contractor in the field.

AESTHETICS:

A transparent rotor unit interferess with the view compared to an uninterrupted expanse of glass. However, the opening, small as it is, provides a strong contact with the outside in terms of admitting the "smell" of fresh air and the sounds of the outdoors. These qualities are largely lost with normal fixed glass windows.

COSTS:

The retail cost of a thru-glass ventilator varies with its size: approximately $7.00 for an 8 inch unit, $5.75 for a 6-1/2 inch unit, and $3.50 for a 5 inch unit. (Prices do not include shipping.) Installation entails simply scribing and tapping out a round hole in the glass. Units are available which lock into place without screws.

EXAMPLES:

The following photograph illustrates one model of a thru-glass ventilator. (76,Simon,p1)

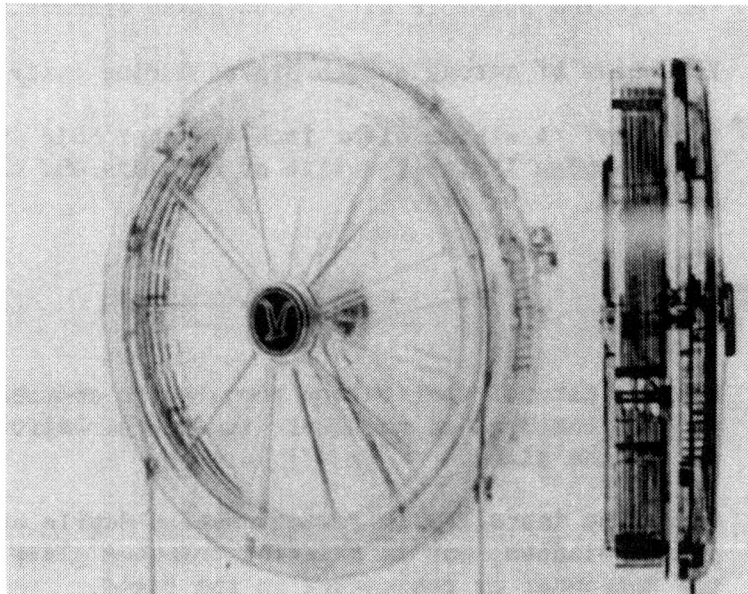

George Roach Co., 8010 24th Ave. NW, Seattle, WA 98107

REFERENCES:

Simon Ltd., "It Runs on Air", George W. Roach Co., Seattle, Wash., 1976.

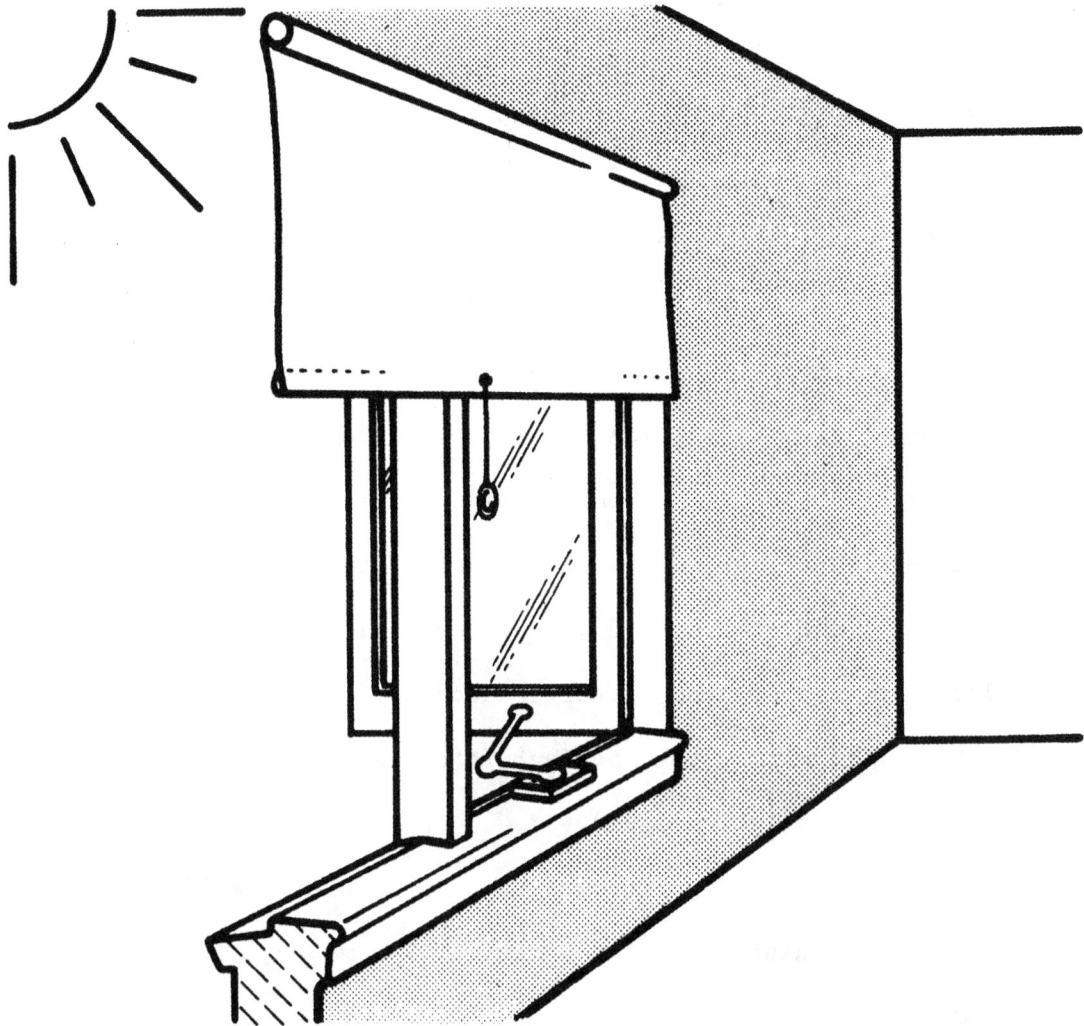

INTERIOR
ACCESSORIES

VENETIAN BLINDS
DRAPERIES
FILM SHADES
OPAQUE ROLL SHADES
INSULATING SHUTTERS

5. INTERIOR ACCESSORIES

The principal advantage of energy-conserving interior accessories
is their accessibility and hence ease of management as outside
conditions change or as the use of the interior changes.
Interior accessories such as draperies, roll shades, and venetian
blinds are effective in reducing heat gain in the summer as
well as reducing heat loss in the winter. The principal disadvantage of interior accessories is the fact that in reducing
heat gain, the heat absorbed in the device is radiated into the
building interior. Also, interior accessories when being used
for shading may limit the opening of in-swing windows when
shading and ventilation are desired. In reducing heat loss, if
the device does not effectively trap air between itself and the
window, the insulative value is minimal. However, if it is
initially installed to provide tight closure and it is subsequently used conscientiously, an interior accessory can greatly
improve the performance of a window.

5.1   VENETIAN BLINDS/Shading, Daylighting

STRATEGY:

Install venetian blinds to reflect the summer sun back out the
window, or to direct daylight to the ceiling for deeper light
penetration into a room.

PHENOMENA:

1)   Slatted horizontal or vertical blinds can be tilted to
provide maximum reflection of sunlight back out the window
in the summer. At a 45 degree tilt with sunlight perpendicular to the slats, blinds have the following properties:
(74,ASHRAE,p403)

Properties of Venetian Blinds

TYPE	TRANS.	REFL.	ABSORBED.
Light-Colored Horizontal	0.05	0.55	0.40
Medium-Colored Horizontal	0.05	0.35	0.60
White (closed) Vertical	0.00	0.77	0.23

These characteristics translate into the following shading
coefficients for blinds in conjunction with different
glazing types: (74,ASHRAE,p402)

SHADING COEFFICIENTS OF VENETIAN BLINDS

TYPE OF GLASS		SOLAR TRANS OF GLASS	MEDIUM HORIZONTAL	LIGHT HORIZONTAL	WHITE VERTICAL[2]
Single	Clear	0.87	0.64	0.55	0.29
Single	Heat-AB	0.46	0.57	0.53	--
Single	Refl.[1]				
	SC=0.30	--	0.25	0.23	--
	=0.40	--	0.33	0.29	--
	=0.50	--	0.42	0.38	--
	=0.60	--	0.50	0.44	--
Double	Clear	--	0.57	0.51	0.25
Double	Heat-AB[3]	--	0.39	0.36	0.22
Double	Reflective				
	SC=0.20	--	0.19	0.18	--
	=0.30	--	0.27	0.26	--
	=0.40	--	0.34	0.33	--

NOTES:

1) Shading coefficients (SC) under the reflective glass column indicate the performance of the glass without interior shading for the purpose of identifying glass types.

2) White vertical blind performance is rated for tightly closed blinds in conjunction with glass having a solar transmittance between 0.71 and 0.80.

3) Heat absorbing glass for outer sheet of glass, clear glass for inner sheet of glass.

2)   Venetian blinds are an effective means of variably controlling the amount of daylight admitted into a room. The
slats can be adjusted to block all direct beam sunlight
while admitting diffuse daylight. With a light-colored
ceiling, the slats can even be tilted to reflect part of
the direct beam sunlight up to the ceiling where it can be
reflected back down to work surfaces. The amount of light
transmitted to the work surface is greatly diminished in
the process but glare is eliminated. Automated control
systems are available to adjust the tilt and even raise
and lower the blinds as the outdoor light level varies.
(See AUTOMATIC SWITCHING/Daylighting)

ADVANTAGES:

1)   Minimal space used to store blinds when open and they
stack with a minimal obstruction of the window area.

2)   Can be selectively tilted to direct daylight to the ceiling
or directly onto the work surface.

3)   Can be partially lowered to eliminate sunlight from only a
portion of a room.

DISADVANTAGES:

1)   Cleaning tedius.

2)   Maintenance of lifting and tilting hardware and cord
replacement.

3)   Decreased effectiveness in reducing winter heat loss
compared to shades or tight-fitting drapes due to cracks
between each slat.

COSTS:

The following are a few sample retail prices for vertical
venetian blinds in the Washington, D. C. area for several
widths and lengths.

	NARROW SLAT 1"	CONVENTIONAL SLAT 2"	VERTICAL BLIND
36 W x 60 H	51.72	18.75	43.00 - 53.00
36 W x 84 H	66.80	26.25	51.00 - 64.50
72 W x 60 H	94.04	37.50	75.50 - 94.00
72 W x 84 H	122.36	52.50	90.00 - 116.50

EXAMPLES:

A study by Arthur Rosenfeld and Stephen Selkowitz proposes the use of reflective venetian blinds in conjunction with clerestory windows to increase the usefulness of beam sunlight in providing illumination. The light entering through high windows is reflected by the blinds up to the ceiling then down to the work surface. It is believed that such a system could provide adequate illumination throughout much of the year. Efficiency could be further improved by coupling the blinds with electric lighting controls. (See Strategy, Automatic Switching.)

The payback period for such a system is illustrated with calculations for a small office of 150 square feet with a 12 foot wide south exposure. A clerestory window the full width of the office would accommodate 12 square feet of beam day-lighting blinds. The clerestory with blinds would deliver an average of 100 lumens per square foot for eight hours on average clear days. Assuming that daylighting could be utilized 80 percent of the occupied hours and the sun shined 65 percent of the time, beam daylight would be viable for 50 percent of the 2,000 annual working hours. (80% x 65%). Assuming 2W/sq. ft. for electric lighting the lighting load is 300W for the office of 150 sq. ft. If the clerestory-blind system provided adequate illumination for half the time, this would then repre-sent a saving of 300 KWH/per year or at 3¢/KWH, $9.00/year. (50 percent of 2,000 hrs x 300W). If the utility company has peak period billing, the savings occur during the peak period and the dollar savings might then be $13/year. A timed light switch is assumed to pay for itself in approximately a year from night and weekend savings. Reflection blind might cost $2.00 per square foot or $24. The payback period for the system is therefore $24/$13 per year or slightly less than two years. (76,Dean,p46)

REFERENCES:

ASHRAE, ASHRAE Handbook of Fundamentals, ASHRAE Inc., New York, N. Y., 1974.

Alcan, "Aluminum Window Blinds for Solar and Light-Control", Alcan Building Products, South Kearny, N. J., 1976.

Avery, "Sundrape Vertical Blind", J. Avery & Co. Ltd., 82-90 Queens land Road, Holloway, London. N7 7AW, April 1976.

Dean, Edward and Rosenfeld, Arthur, Efficient Use of Energy in Buildings, LBL 441, Lawrence Berkeley Laboratories, Berkeley, Calif., 1976.

Dix, Rollin C., and Zalman, Lavan, "Window Shades and Energy Conser vation". Illinois Institute of Technology, Chicago, Ill., Dec. 1974.

Levolor, "Window Magic", Levolor Lorentien Inc. Hoboken, N. J., 1976.

Marathon Carey-McFall, "Bali-Architect Venetian Blinds", Marathon Carey-McFall Co., Philadelphia, Pa., 1974.

Stephenson, D. G. and Mitalas, G. P., "Solar Transmission through Windows with Venetian Blinds", Research Paper No. 310, Division of Building Research, National Research Council, Ottawa, Canada, April 1967.

5.2 DRAPERIES/Shading, Insulation

STRATEGY:

Install draperies sealed against the wall or window frame at the sides and extending down in contact with the floor or window sill to insulate the window in the winter and provide shade in the summer.

PHENOMENA:

1) Heat loss through windows with tight fitting closed draperies is substantially reduced compared to the heat loss of an uncovered window. The effectiveness of a closed drapery as an insulator is greatly impaired if conditioned air is free to circulate between the drapery and the window. When room air comes in contact with the cold glass, it is cooled and cascades back into the room at the bottom of the drapery. Under such conditions, the winter U-value of a single glazed window is only reduced from 1.13 to 1.06. (74,Dix,p13). By contrast, the winter U-value of a tight-fitting, tight-weave closed drapery and single glazed window can be assumed to be as low as 0.88. (calculated from 74,ASHRAE,p395)

2) In the winter the draperies should be opened when the window is sunlit to allow the sunlight to penetrate into the room, warming more massive materials and remote surfaces. The heat will then radiate to other interior surfaces rather than directly back to the glass. The drapery track should extend a sufficient distance to either side of the window to permit the draperies to stack clear of the window to allow all the sunlight to penetrate into the room.

3) Comfort near windows with drawn draperies will be improved compared to uncovered windows. This is due to the draperies being much closer to room temperature than the glass with a corresponding reduction in body heat loss by radiation. Tight-weave fabrics are more effective for improving comfort.

4) Summer heat gain can be reduced with draperies. The effectiveness of the drapery is mainly determined by three factors: the amount of incoming solar energy reflected back at the glass, the amount of solar energy absorbed by the fabric, and the amount of solar energy transmitted through the fabric and through the openings of the weave. To a lesser extent, the insulation value of the drapery also affects how much outdoor heat is added to the air conditioning load. For single glazed windows with tight fitting draperies, the summer U-value can be as low as 0.81 compared to 1.06 for the uncovered window. (74, ASHRAE,p395) The shading coefficient of single glazed windows (1/4" plate) with draperies ranges from 0.80 down to 0.35, the low value representing a highly reflective tightly woven drapery material. (74,ASHRAE,p405). The window heat gain can be calculated by adding the solar heat gain and the conducted heat gain. The solar heat gain equals the amount of solar energy transmitted plus the amount of solar energy absorbed by the configuration and dissipated into the building interior. The conducted heat gain equals the U-value multiplied by the inside outside temperature difference. (See 74,ASHRAE,p388)

5) The use of tight fitting draperies with insulating glass even further reduces the summer U-value as shown below: (74,ASHRAE,p395)

DOUBLE GLASS (width of air space)	NO SHADING	WITH INTERIOR SHADING
single glazing	1.06	0.81
3/16" air space	0.66	0.54
1/4" air space	0.65	0.52
1/2" air space	0.59	0.48

6) The use of double draperies, two layers of draperies separated by an air space, further improves the thermal performance of windows. A summer U-value of 0.65 is possible for single glazed windows assuming the same degree of air tightness for both layers of trapped air. Assuming a better degree of air tightness between the two layers of drapery than between the drapery and the window results in the even lower calculated U-value of 0.57. (Pennington, p2)

7) Draperies should be installed such that conditioned air blows on the room side of the drapery and not between the drapery and the window. Where the register is directly below or above the window, retrofit deflectors are readily available which divert the air into the room rather than up or down the window surface.

ADVANTAGES:

1) Decreased winter heat loss and summer heat gain.

2) Improved comfort near window possible when draperies closed.

3) Glare control.

4) Privacy

5) Noise absorption. Noise within a room is absorbed by draperies rather than reflected as from uncovered glass. Also, outside noise transmitted through the glass is partially absorbed. The denser the weave and heavier the drapery, the more effective it is in reducing noise transmission. The following graph illustrates the effect of the openness of the weave on sound reduction. (74, ASHRAE,p406)

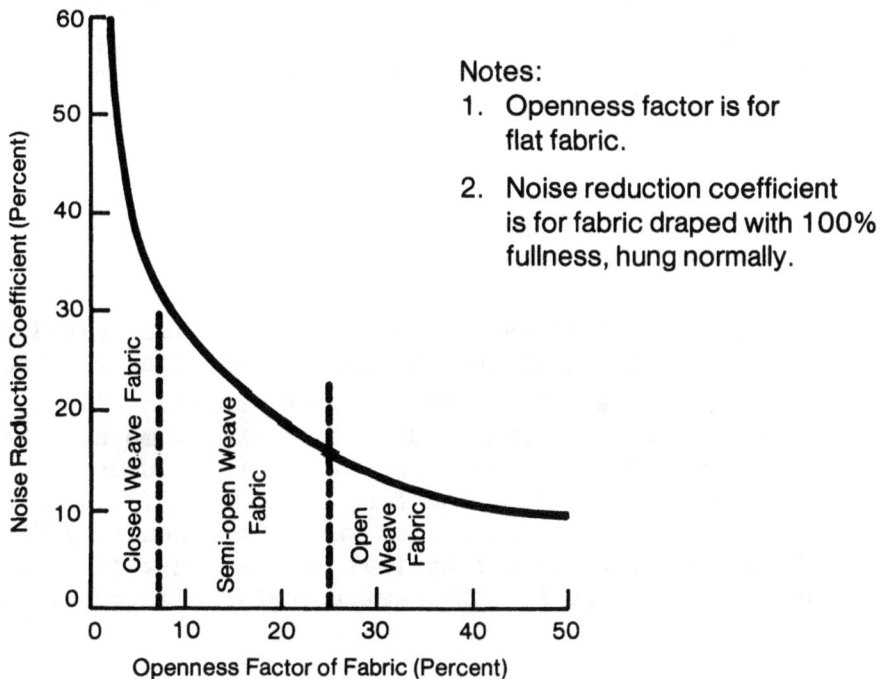

Figure 37   Drapery Type vs. Noise Reduction

DISADVANTAGES:

1) Periodic cleaning required.

2) Obstruction of view when closed.

3) Possible breakage of glass when used in conjunction with heat-absorbing glass. (Also true for any reflective interior shading device.) The glass and the way it is set should be designed to withstand the additional heat-build-up from sunlight being reflected back at the glass from the closed drapery.

AESTHETICS:

The desire for an airy open weave drapery conflicts with the thermal effectiveness of draperies both summer and winter.

COSTS:

The price ranges largely with the cost of the fabric. A ready-made drapery can range in price from $0.85 to $1.50 per square foot. Custom made draperies may range much higher in price. The traverse rod is frequently available in various length increments which are expandable to any size up to the next size increment. A 12 to 15 foot traverse rod can range from $12 to $18.

EXAMPLES:

A series of tests were conducted by the Illinois Institute of Technology to determine the percentage of energy that could be saved by using shades, drapery, or venetian blinds. The tests were conducted in two rooms with a window mounted between them. One room had its temperature varied from 20 to 50°F (-6.7 to 10°C) for winter test and from 85 to 95°F (29.4 to 35°C) for summer test. The other room was maintained at 75°F (23.9°C). Solar radiation levels typical of the mid-west were introduced using heat lamps for the summer test. All cracks were sealed around the window to prevent infiltration. The draperies were hung two inches out from the wall so they completely covered the window opening. The study concluded that a medium colored drapery with a white plastic backing reduced conducted heat loss in the winter by 6 to 7% (74,Dix,pii) and conductive and radiant heat gains in the summer by 33% (74,Dix,p21)

REFERENCES:

ASHRAE, ASHRAE Handbook of Fundamentals, American Society of Heating, Refrigeration, and Air Conditioning Engineers, New York, 1974.

Dix, Rollin C. and Zalman, Lawan, "Window Shades and Energy Conservation", Mechanics Mechanical and Aerospace Engineering Dept., Illinois Institute of Technology, Chicago, Ill., Dec. 1974.

Keys, M. W. "Analysis and Rating of Drapery Materials Used for Indoor Shading", ASHRAE Transactions, Vol. 73, Part I, ASHRAE Inc., New York, N. Y., 1967.

Morrison, Clayton A. et.al, "An Experimental Determination of Shading Coefficients for Selected Insulating Reflective Glasses and Draperies", No. 2382 R132, Mechanical Engineering Dept., University of Florida, Gainesville, Fla., undated.

Ozisik, N. and Schutrum, L. F., "Solar Heat Gain Factors for Windows with Drapes", ASHRAE Transactions, Vol. 66, ASHRAE Inc., New York, N. Y., 1960.

PA, "Window Coverings", Progressive Architecture, Reinhold Publishing Co., Stamford, Conn., Nov. 1966.

Pennington, C. W., et.al, "Experimental Analysis of Solar Heat Gain Through Insulating Glass with Indoor Shading", ASHRAE Journal, ASHRAE Inc., New York, N. Y., Feb. 1964.

Pennington, C. W., et.al, "Analysis of Double Drape Fenestration Configurations", Dept. of Mechanical Engineering, Univ. of Florida, Gainesville, Fla.

PPG, "The Feneshield System", PPG Industries, Pittsburgh, Pa., Jan. 1974.

Yellott, J. I., "Drapery Fabrics and Their Effectiveness in Sun Control", ASHRAE Transactions, Vol. 71, Part 1, ASHRAE Inc., New York, N. Y., 1965.

5.3 FILM SHADES/Insulation, Solar Heating, Shading

STRATEGY:

Install clear or coated transparent film shades--singly or in separated multiple layers and sealed at the sides and bottom to provide insulation, improved air tightness, solar heating, and/or shading.

PHENOMENA:

1) A roll shade sealed against the perimeter of a window frame will create an insulating layer of air. This effect can be multiplied by providing several consecutive layers of shades and air spaces. One such system is reported to provide a U-value for the window of 0.55 with one shade pulled down, 0.31 with two shades pulled down, and 0.18 with three shades pulled down. (76,Insealshaid,p4)

2) A roll shade sealed at the edges also impedes infiltration.

3) A heat-absorbing film shade in sunlight can have a surface temperature above room temperature and thereby provide a warm radiant surface improving comfort.

4) A low-emissivity film shade can reduce the window's absorption of heat radiated from interior wall surfaces, furniture and people. This can result in a heat-loss reduction of 57 to 64 percent from single glazed windows without a shade. (75,Dahlen,"Summary"p2)

5) A reflective film shade can reflect as much as 60 percent of the incoming sunlight back out through the window. (76,Joanna,p2)

6) A selective transmissivity film shade can transmit 75 percent of the visible light but only 55 percent of the total solar radiation. This allows usable daylight to enter but blocks much of the invisible radiation, thereby reducing heat gain. (76,Doyle)

ADVANTAGES:

1) Reduced summer solar heat gain and winter night losses
    with a reflective or selective film, singly or in com-
    bination with clear film.

2) Reduced winter heat loss with shade configurations which
    provide insulating, trapped-air space(s).

3) Reduced infiltration provided by obstructing flow of
    incoming air with a film shade sealed at the edges.

4) Increased comfort near windows because less body heat is
    radiated to an inside film shade than to a cold window.

5) Elimination, with a heat-absorbing film, of ultra-violet
    radiation and its resulting fading of carpets and furniture

6) Reduced sky glare with heat-absorbing films.

7) Night visibility into building for security surveillance,
    (e.g., shops, schools, or banks).

8) Daytime privacy with reflective film shades.

9) Self-storing when not in use.

DISADVANTAGES:

1) Required management of shade by occupants of building.

2) Difficult to install on other than right angled window
    areas.

3) Lack of glare control from direct sunlight.

4) Lack of privacy with clear shades or reflective shades at
    night.

5) Reduced winter solar heat gain through low emissivity film
    shades installed to reduce heat losses. Calculation of
    the net heat gain or loss for a given location may suggest
    that low emissivity film shades are effective on north,
    east, and west orientations and transparent uncoated film
    shades are effective on south orientations.

AESTHETICS:

Film shades may become unattractive after several years due to scratches and wrinkles, especial at the edge of the shade. However, replacement of the film is substantially less expensive than the initial installation of the complete system.

COST:

Delivered Film Cost:

Reflecting Film (3m)	20¢/sq.ft.
Clear Film	18¢/sq.ft.
Absorbing Film	30¢/sq.ft.

A light filtering or reflective film shade mounted on a roller costs approximately:

3 x 5 ft. window $32.50 ($2.16/sq.ft.)
5 x 6/8 patio door $57.50 = ($1.72/sq.ft.)

The estimated installed cost for single-layer shade system of either a clear or a reflecting film with a magnetic edge seal is:

$1.50/sq.ft. on large job (e.g., 100 windows)
$2.00/sq.ft. on small job (e.g., 10 windows)

It may be possible to develop a do-it-yourself magnetically sealed shade system for the homeowner at a cost of less than $1.00 per square foot.

The estimated installed cost of a multiple shade system consisting of clear film, heat-absorbing film, and reflective film with sealed edges and automatic venting is:

$4.00/sq.ft. for large windows
$6.00/sq.ft. for small windows

EXAMPLES:

1) One company manufactures a low-emissivity film shade with
a magnetic edge tape for sealing the perimeter to an iron
oxide tape adhered to the jamb and sill. Test were conducted on a school in Minneapolis on November 24, 1974.
Measurements were made, using an infrared temperature
sensor and thermocouple, on five adjacent windows on the
first floor classroom. One window was uninsulated while
the other four received various retrofit options.

Infrared photographs were then taken on December 1, 1975
to confirm the measurements. The tests showed the insulating effectiveness of the various windows as follows:
(75,Dahlen,p2)

<u>Percent Heat Loss Reduction</u>
(comparison to single glazed windows)

(1) Conventional roll shade plus a   28-36%
    blackout shade for movie pro-
    jecting.
(2) Clear plastic film shade with   36-43%
    all perimeter sealed.
(3) Wooden frame exterior storm window.   50-57%
(4) Low heat emitting film shade sealed   57-64%
    on bottom and sides only.

2) Another company manufacturers a system of three shades: a
reflective film shade near the window, a heat-absorbing
film shade near the room, and a clear shade in the middle.
The shades are operated within frame guides to seal the
sides and bottom.

In the winter during the day, the occupant should lower
the clear shade and the heat-absorbing shade. The clear
shade creates an insulating air space between itself and
the window. The heat-absorbing shade heats the air in
the space between itself and the clear shade. When the
trapped air temperature becomes warm enough, a bi-metalic
thermostat opens flaps covering slots at the bottom of
the shade frame. Room air is then drawn in at the bottom,
heated by the warm, heat-absorbing film surface, and
convected back into the room at the top. In the winter
during the night, all three shades are lowered to provide
three insulating, dead-air spaces.

In the summer during the day, the outer reflective film
is pulled down to reflect 80% of the sunlight and the
middle clear film is pulled down to provide insulation.
(76,Insealshaid,p6)

The Sectional drawings following the references illustrate combinations of roll shades possible with such a
system.

REFERENCES:

Dahlen, R. R., Doyle, J. S., Klaenhamma, B. L., "Summary of Window
    Insulation Tests for Double-Hung School Windows." 3M Company,
    St. Paul, Minneapolis, December 1975.

Dahlen, R. R. "Laboratory Measurements of Window Shade Thermal
    Performance", 3M Company, St. Paul, Minneapolis.

Dahlen, R. R. "Window Air Infiltration Measurements in School
    Classroom", 3M Company, St. Paul, Minneapolis.

Doyle, James S. and Dahlen, R. R. Meeting at NBS.

Insealshaid, "The More Windows You Have, the More You Need Insealshaid
    Ark-tic-seal Systems Inc. Butler, Wis., 1976.

Joanna Western Mills, "The Money Saving Way to Daytime Comforts",
    Joanna Western Mills Co., Chicago, Ill., 1976.

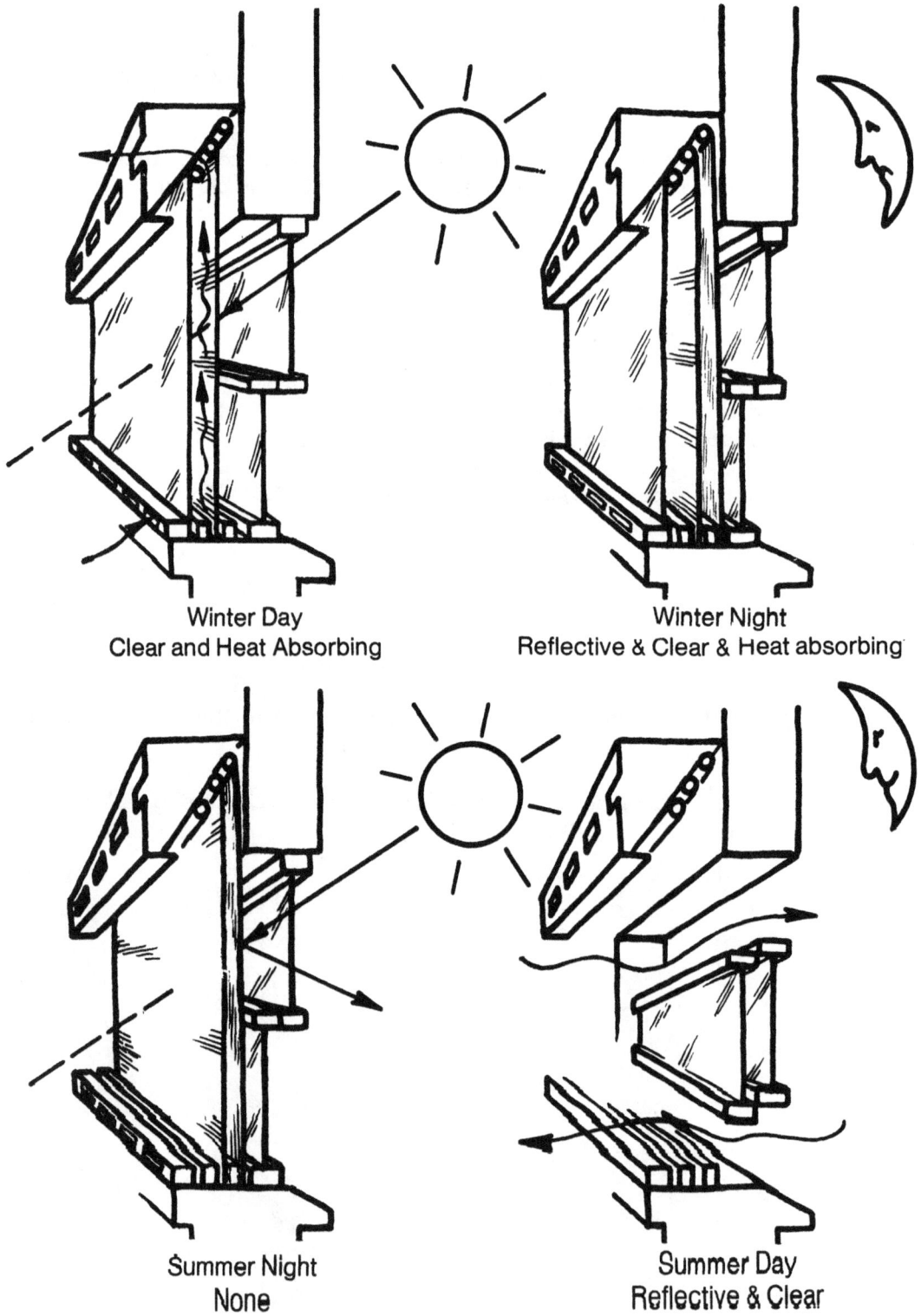

Winter Day
Clear and Heat Absorbing

Winter Night
Reflective & Clear & Heat absorbing

Summer Night
None

Summer Day
Reflective & Clear

Figure 38. Use Modes of a Multiple
Shade System

STRATEGY:

Install an opaque or translucent roll shade to reflect sun
back out the window in the summer, and, if the shade has a
dark side and is reversible, absorb solar energy in the
winter.

PHENOMENA:

1)   Summer solar heat gain through windows can be reduced by
      lowering an opaque, white roll shade to reflect much of
      the incoming solar energy back out through the glass.
      The color of the shade and its opacity greatly affect
      performance as can be seen in the following table:
      (74,ASHRAE,p403)

CHARACTERISTIC	TRANSMITTED	REFLECTED	ABSORBED
Light-color, translucent	25%	60%	15%
White, opaque	0	80	20
Dark, opaque	0	12	88

As is evident from the table, a shade's ability to reflect
sunlight is badly impaired if it is a dark color.

The light absorbed by the shade raises the shade temperature
Heat is then dissipated into the room by radiation to
room surfaces, and by convection of room air in contact
with the warm shade surface.

2)   The effectiveness of roll shades in combination with
      various types of glass is shown below in terms of shading
      coefficients: (74,ASHRAE,p403)

GLASS	SOLAR TRANSM.	SHADING COEFFICIENT		
		OPAQUE DARK	OPAQUE WHITE	TRANSLUC LIGHT
Single Clear	87%	0.59	0.25	0.39
Single Heat-Ab.	35*	0.45	0.30	0.36
Double Clear	69*	0.60	0.25	0.37
Double Heat-Ab.	28*	0.40	0.22	0.30

* (75,LOF,p19)

3) Roll shades also reduce heat flow through a window both in winter and summer. The U-value of a roll shade with a moderately close fit to the window opening in the wall will achieve a summer U-value of approximately 0.88. (Calculated from 74,ASHRAE,p395)

4) A roll shade can have a dark color on one side which effectively absorbs sunlight and a white surface on the reverse side which effectively reflects sunlight. By simply reversing the shade from dark side facing out in winter to reflective side facing out in summer, the shade can perform as a solar collector or shading device varying with the season. (Silverstein,p63)

ADVANTAGES:

1) Reduced solar heat gain in summer.

2) Reduced conducted heat loss during winter nights.

3) Privacy.

4) Glare control.

5) Self-storing.

DISADVANTAGES:

1) Maintenance. Spring mechanism will fatigue or jamb with time and need replacement.

2) View out and shading at the same time impossible.

3) Impeded ventilation from the top opening of double hung windows when the shade is lowered.

AESTHETICS:

A roll shade is a simple, inobtrusive element which does not complicate the appearance of a window.

COSTS:

The following are a sample of retail prices for roll shades in the Washington, D. C. area:

SHADE WIDTH	TYPE OF SHADE	
	OPAQUE	TRANSLUCENT
36 inches	$5.00	$3.70
60 inches	20.00	17.00

EXAMPLES:

The following example illustrates the summer and winter energy benefit or expenditure attributable for one square foot of window without and with a roll shade in New York City. The summer calculated energy results plus the solar data, degree days, and window U-values are taken from a report by Samuel Berman. (75,Berman,p47) The winter energy data and total energy data are not from Berman. These calculations differ from his report in order to consider the window with a lowered roll shade during winter nights. Three orientations are calculated with west being assumed similar to east.

The following values are used in the calculation:

Type of Glass	Winter U-Values	
	Shade Up	Shade Down
Single glass	1.13	0.88
Storm/double	0.55	0.49

Type of Glass	Winter Solar Heat Gain (WSHG) (KBTU/winter, sq.ft.)		
	N	E,W	S
single	45	91	159
	39.6		139.9
double		80.1	

The shading coefficient for double glass is assumed to be 0.88.

Heat loss for the winter was calculated with the shade lowered 12 hours per day. The heating season used was 4714 degree days (d) for a period from October to April.

$$q = (WSHG \times SC) - (12 \times U_{shade} \times d) + (12 \times U_{no\ shade} \times d)$$

Using these assumptions the following winter and summer energy expenditures or benefits result from one square foot of window.

SEASONAL ENERGY EXPENDITURE FOR A WINDOW IN NEW YORK CITY (KBTU/SQ.FT.) (- denotes energy input required from mechanical system)

GLAZING	SHADING	NORTH		EAST		SOUTH	
		Wint.	Sumr.	Wint.	Sumr.	Wint.	Sumr.
Single	None	−84	−43	−38	−76	+29	−59
Single	roll shade	−69	−15	−23	−26	+45	−20
	Savings	15	28	15	50	16	39
Storm/double	None	−25	−37	+14	−65	+71	−51
Storm/double	roll shade	−19	−13	+21	−23	+81	−18
	Savings	6	24	7	42	10	33
Savings roll shade + double compared to no roll shade + single		65	30	17	53	52	41

REFERENCES:

ASHRAE, ASHRAE Handbook of Fundamentals, ASHRAE Inc., New York,
    N. Y., 1974.

Berman, Samuel and Silverstein, Seth, "Energy Conservation and
    Window Systems" National Technical Information Service, Spring-
    field, Va. Jan. 1975.

Dix, Rollin C. and Lavan Zalman, "Window Shades and Energy Con-
    servation" Illinois Institute of Technology, Chicago, Ill., 1974.

Land, C. E., "Effect of Reflective Window Shades in Reducing the
    Heat Transfer through Windows", University of Minnesota, Minne-
    apolis, Minn., 1957.

LOF, "Glass for Construction", Libbey-Owens Ford Co., Toledo,
    Ohio, 1975.

Olgay, Victor and Olgyay, Aladar, "Design with Climate", Princeton
    University Press, Princeton, N. J., 1963.

Ozisik, N. and L. F. Schutrum, "Heat Flow through Glass with Roller
    Shades", No. 1696, ASHRAE Transactions, Vol. 65, ASHRAE Inc.,
    New York, N. Y., 1959.

Silverstein, Seth B., "A Dual-Made Internal Window Management Device
    for Energy Conservation", General Electric Co., Schenectady,
    N. Y., undated.

5.5   INSULATION SHUTTERS/Insulation

STRATEGY:

Provide hinged or removable opaque insulating shutters to
reduce night-time winter heat loss.

PHENOMENA:

1)   The winter heat loss through a window can be reduced by
      covering the window with an insulating panel in contact
      with the glass. The heat loss is then reduced in pro-
      portion to the insulating value of the panel measured as
      resistance to heat flow per inch thickness of material.
      The U-value for a window with a one inch insulating panel
      against it can be approximated as follows:

$$U_{total} = (\frac{1}{U_{glass\ alone}} + R_{panel})^{-1}$$

This assumes equivalent values for the air film at the
surface of the panel and at the interior surface of the
glass, or an R-value of 0.68 in both cases.

The following table lists several common rigid insulating
materials, their resistances (74,ASHRAE,p361), and the
winter U-value of a single glazed window with the panel
in contact with the glass. All values are for one inch
thick panels.

INSULATING VALUES OF VARIOUS INSULATING PANEL MATERIALS

Material	Resistance (for 1" thickness)	U-value of window with 1" panel	U-value of window with 2" panel
Expanded polystyrene, extruded, plain	4.00	0.20	0.11
Expanded polystyrene, molded beads	3.57	0.22	0.12
Expanded polyurethane	6.25	0.14	0.07
Cork (3/4 inch)	1.68[1]	0.39[1]	
Cork/paper bd/cork (3/4)	2.56[1]	0.29[1]	--
Plywood (3/4 inch)	0.93	0.55	--

2) If a gap exists between the insulating panel and the glass and air can circulate from the room behind the panel and across the glass, the effectiveness of the panel in reducing heat loss will be drastically reduced. However, if the panel fits tightly to the perimeter of the wall opening (so that air cannot circulate into the room), a separation of the panel from the glass will increase the panel's effectiveness by providing an insulating layer of trapped air.

ADVANTAGES:

1) Reduced winter heat loss at night.

2) Improved comfort near windows at night.

3) Reduced sound transmission at night.

4) Privacy at night.

DISADVANTAGES:

1)   Bulky to store during the summer and during day if
     demountable panels are used.

2)   Easily crushed in the case of polystyrene or polyurethane
     necessitating periodic replacement.

3)   Polystyrenes and polyurethanes are flamable and give off
     highly toxic gases in the event of fire. A protective
     cladding of metal is essential.

AESTHETICS:

Rigid insulating panels can be covered with decorator fabrics
to enhance the character of the window. Cork or other insulating materials are attractive in their natural condition in
many interior design schemes. In deep window openings, the
window may be sufficiently recessed to accommodate half the
width of the window, in which case, thermal panels can be
pivoted at either side without projecting out into the room.
In new construction a pocket can be detailed in the wall
adjacent to the window into which sliding insulating panels
can be stored out of sight. Foam panels should be clad
with sheet metal. The metal skin would eliminate the
problem of panels becoming unsightly due to the vulnerability
of unprotected foam. The skin must also provide protection
from the toxic fumes given off by certain types of foam
insulation in the presence of a fire.

COSTS:

The following table is a sample of retail prices for various
insulating materials:

Material	Cost/sq. ft.
1" extruded polystyrene	$0.35
1" expanded polystyrene	0.38
3/4" sheet cork	0.50
1/2" pressed paper board with 1/8" cork facing both sides	1.51

EXAMPLES:

1) An insulating panel which reduces a single glazed window's U-value to 0.38 or less, is twice as insulating as residential double glass (1/4 inch air space) with a U-value of 0.65. Thus, if the panels are closed only during the hours of darkness, approximately twelve hours per day, they will achieve energy savings comparable to insulating glass. Furthermore, the difference between outside day and night temperatures is likely to be greater than the difference between day and night thermostat settings. Therefore, even with night-time thermostat set back the insulating shutters can effectively conserve heating energy.

2) A study in Sweden calculated the difference in energy flow between insulated glass windows with thermal shutters open all the time and with the shutters closed when conductivity and infiltration losses exceed solar gain. The following table gives the annual energy saved by using the shutters in BTUs x 10^3 per square foot. (Converted to English Units from 75,Hagman,p2)

LOCATION	DEGREE DAYS	NORTH	ORIENTATION SOUTH	EAST/WEST
Lulea	11000	142	125	135
Stockholm	7700	106	89	99
Malmo	6900	93	79	87

NOTES:

1) The case of closing the shutters whenever the window loses more heat than it gains would necessitate closing the shutters the entire day during December in Malmo and Stockholm, and November through January in Lulea.

2) The window with a closed shutter was assumed to have an equivalent U-value of 0.09 considering both conduction and infiltration. With the shutter open, the window alone, considered to be an average tightness, double glazed unit, the U-value assumed to be equivalent to 0.53.

REFERENCES:

ASHRAE, ASHRAE Handbook of Fundamentals, ASHRAE Inc., New York, N. Y., 1974.

Hagman, Folke, "The Window as an Energy Factor, Insulating Shutters, Function Construction, Economy", Report R43: 1975, National Swedish Council for Building Research, S-111 84 Stockholm, Sweden, 1974.

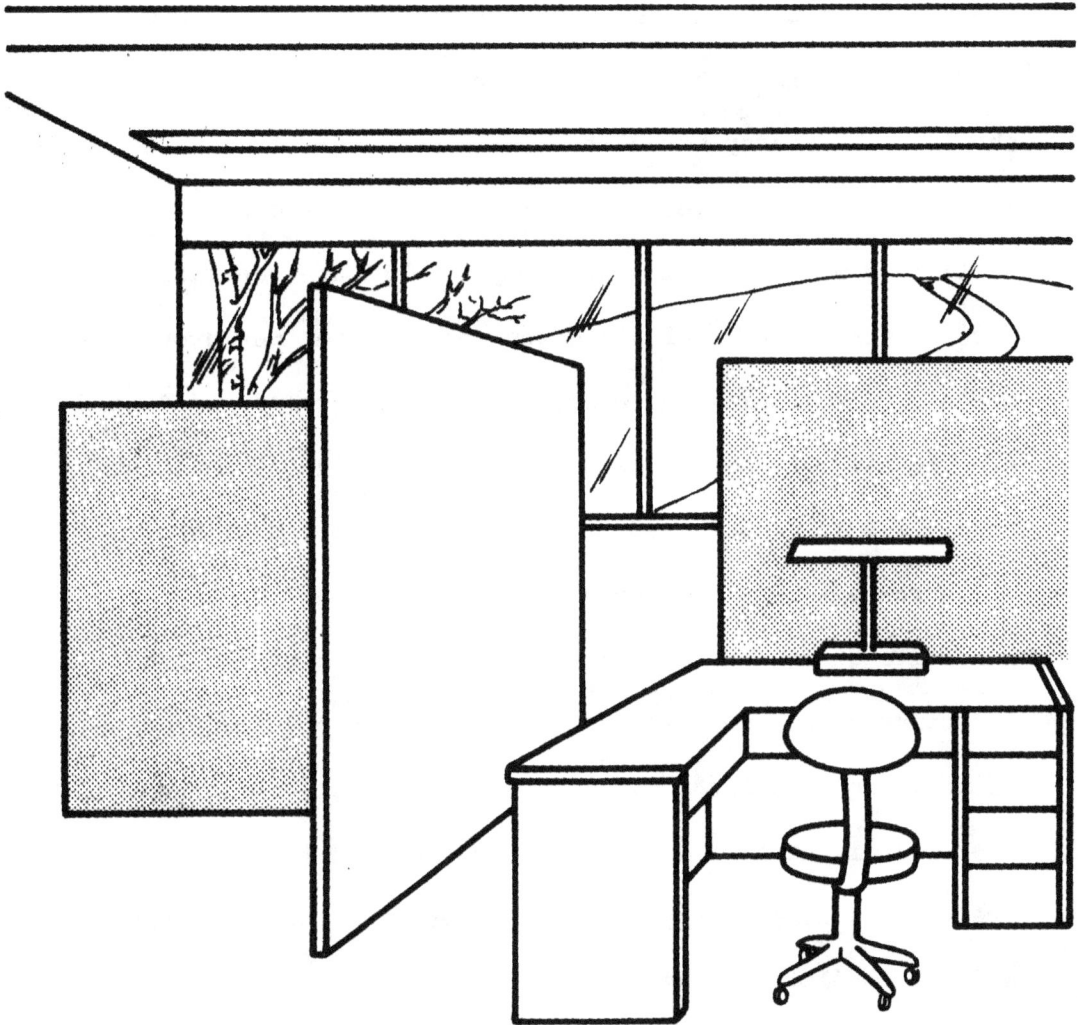

BUILDING
INTERIOR

FIXTURE CIRCUITING
TASK LIGHTING
AUTOMATIC SWITCHING
INTERIOR COLORS
THERMAL MASS

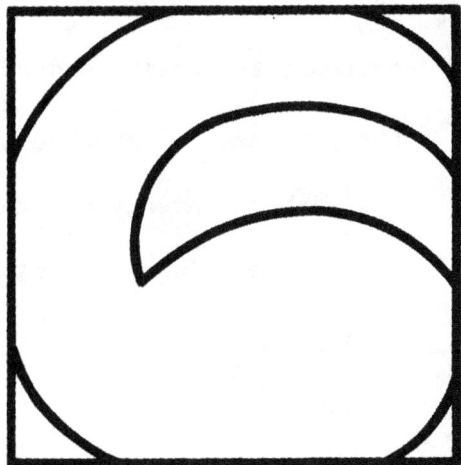

6. BUILDING INTERIOR

The design of the building interior will determine how useful
the energy admitted by the window will be. Distribution of
the incoming energy beyond the immediate vicinity of a window
is one basic objective. The color, location and height of
partitions will determine how deeply daylight penetrates into
a room. The mechanical air circulation system may provide a
means of circulating the winter solar heat beyond the immediate vicinity of the window. Another objective of interior
strategies is to store solar heat during periods of excess
to be re-radiated later when the room temperature drops.
A massive floor and wall in the path of the sunlight can
effectively perform this function. Still another objective
of building interior strategies is to facilitate occupant's
use of daylight as a substitute for electric lighting,
thereby reducing energy consumption by both the lighting
and air conditioning systems.

In summary, the design of the illumination and mechanical
systems, the design and color selection of room surfaces, and
the placement of massive building components will determine
how much the energy gained by windows can reduce illumination,
air conditioning, and heating energy costs.

STRATEGY:

Provide separately switched circuits for lights in a building's peripheral zone to facilitate substituting daylighting for electric lighting.

PHENOMENA:

1) Daylight penetrates a finite distance into an interior space. The illumination level can be calculated and a zone defined where daylight provides an acceptable level of illumination with no artificial lighting needed.

2) The amount of daylight penetrating a given space varies according to cloud cover, time of day, and time of year, as well as the visible light transmission of the glazing. Providing separate switching for two rows of light, one near the window and one deeper in the room, would permit the outer row to be switched off when daylight is adequate near the window.

3) Daylighting, when excesses are controlled, poses less of a heat load on the air conditioning system than even fluorescent lighting. Therefore, the capability of switching on only the row of lighting needed on a hot overcast day saves electricity both in terms of lighting and air conditioning.

ADVANTAGES:

1) Reduced electric consumption during the time of the day when demand on the generating plant is at a peak. Where electric rates are higher during peak periods, dollar savings are even greater.

DISADVANTAGES:

1) Potential glare and greater variation in light level compared to complete reliance on artificial illumination.

2) Overheating if excesses not controlled with shading.

AESTHETICS:

1) Turning off only the light fixtures in the area receiving daylighting may provide a more uniform lighting level throughout the space.

COST:

Additional first costs include the material and installation of additional footage of wiring and additional switches.

EXAMPLES:

1) The following figure illustrates the amount of light theoretically available from a window in an office space. The clear sky curve assumes the sun is directly overhead. If the sun were lower in the sky as in the winter, the light level in the office would be still higher than shown. If the sun were behind the building, the light level would approach the curve for the overcast sky. (74,Vild, p3) The actual illumination in any office may be substantially lower than these curves. Lower values would actually be better as the illumination levels shown are excessive.

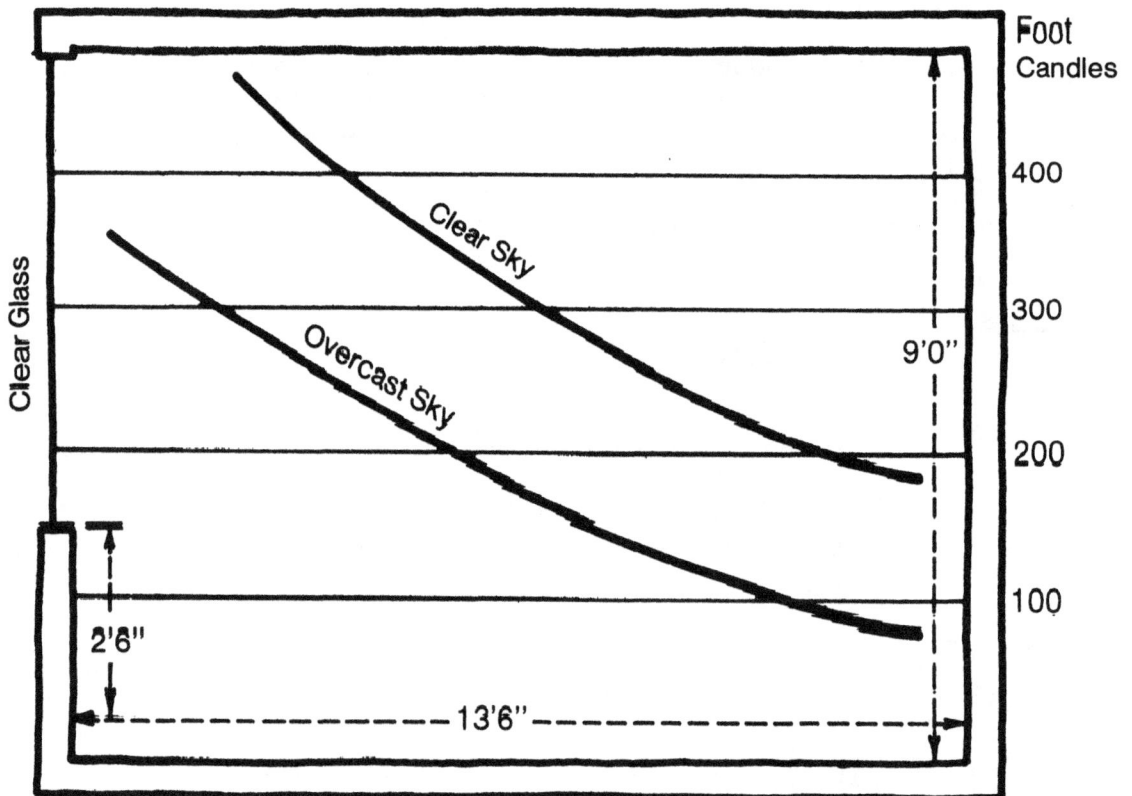

Figure 39. Usable Daylight at Desk- Top Height vs. Distance from the Window

6-2

2) The following example shows how much electricity can be
   conserved when daylight would suffice to provide adequate
   illumination for half the office six hours a day.

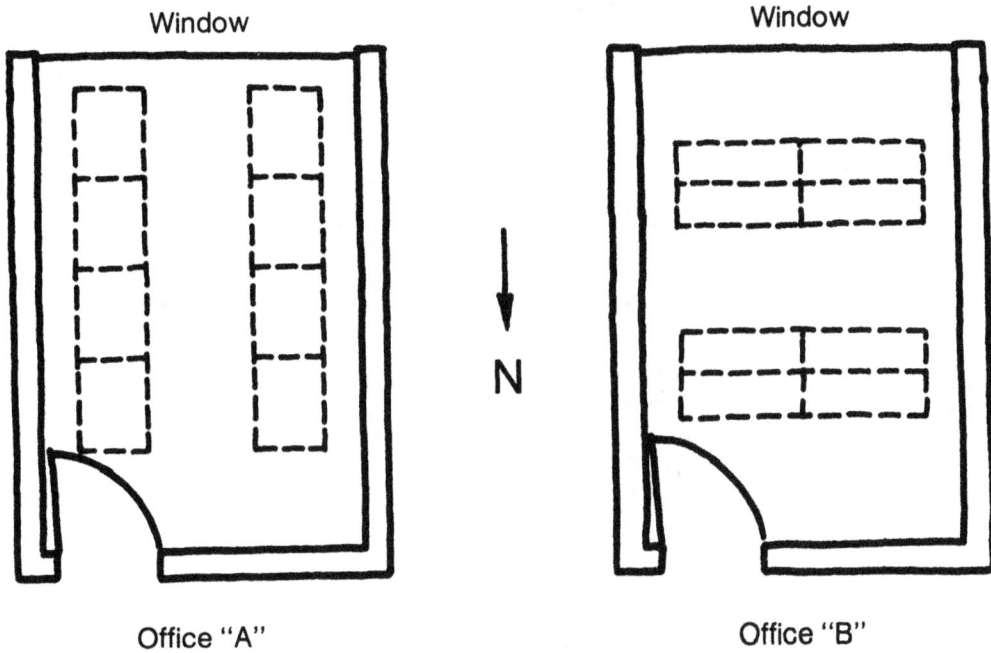

Figure 40. Lighting Fixture Circuiting

8 fixtures
2 - 40W tubes each

Assume:

| 8 fixtures | 4 fixtures | 4 fixtures |
| 9 hrs/day | 9 hrs/day | 3 hrs/day |

5.76 KWH/work day   2.88 KWH/work day + 0.96 KWH/work day = 3.84

$1.15/work week $0.77/work week

    Savings = $0.38/work week

REFERENCES:

Vild, Don, "Benefits of Daylighting in Terms of Energy Savings",
    Libbey-Owens Ford Corp., Toledo, Ohio, 1974.

STRATEGY:

Provide task lighting switched separately from ambient lighting to facilitate the use of daylighting for ambient lighting.

PHENOMENA:

1)   Ambient lighting in addition to adequate lighting on a
      task is necessary to avoid extreme contrast and resulting
      eye fatigue. However, the ambient lighting can be only
      approximately 1/3 the level of the task lighting.
      (75,GSA,p6-6) By providing high levels of lighting only
      at the task rather than uniformly high levels of light,
      considerable electricity can be saved, both from reduced
      lighting consumption and reduced air conditioning burden.
      By providing separate switching for task and ambient
      lighting, the ambient lighting can be turned off when
      daylight is adequate, further increasing the savings.

ADVANTAGES:

1)   Reduced electricity consumption when daylighting is
      adequate for background lighting.

2)   Individual discretion in selecting lighting for a task.
      With conventional uniform lighting, if one person feels
      he needs more light he must switch on the lighting for
      everyone.

3)   No electricity, including the resulting increase in air
      conditioning load, is wasted by lighting work areas of
      individuals who are absent for part or all of the day.

4)   Depending on the mounting system an individual can vary
      the distance and orientation of the luminaire, relative
      to his task, to suit his personal preference.

DISADVANTAGES:

1)   Possible glare from the task lighting reflected off the
      work surface into the eyes of the worker.

2)   Interference with work task by fixture base if the task
       lighting fixtures are on the work table.

3)   Less flexibility in terms of relocating the task at a
       future date if task lighting fixtures are mounted in the
       ceiling.

AESTHETICS:

1)   In illuminating the task area more brightly than the
       background, attention is focused on the work surface.

2)   Introducing different levels of illumination within a
       space makes it more interesting than if it were lighted
       uniformly.

3)   Task luminaires can be an attractive addition to the
       furnishing of a space.

COSTS:

An incandescent drafting lamp can be purchased for $19.00.

EXAMPLES:

The office of the Pilkington Environmental Advisory Service;
Lancashire, England was experimentally retrofitted with task
lighting and daylighting was relied upon to provide back-
ground lighting as much as possible. This was accomplished
by providing desk lamps or drafting lamps and removing nearly
three-quarters of the fluorescent lighting tubes selectively
from the ceiling. Desks were faced towards the darker center
of the space with daylight then coming from behind the desk
rather than from in front of it. This reduced "veiling
reflection". (Glare from light reflected by the pages reducing
the contrast between the ink and the paper.) The following
table summarizes the lighting wattages in the original overall illumination system and the retrofitted task lighting and
daylight ambient lighting system. (75,Cuttle,p24)

COMPARISON OF ORIGINAL AND RETROFITTED LIGHTING

Installation	Original	Retrofitted
Total lighting (W)	5130	2866
Lighting load W/m^2	24.4	13.6
Density of occupancy (m^2/person)	15	15
Background lighting load (W)	--	95
Total lighting load/desk worker (W)	365	149
Total lighting load/illustrator (W)	365	205

REFERENCES:

Architectural Record, "Task/Ambient Lighting, An Idea Whose Time Has Come, But Whose Implications Need to Be Better Understood", Architectural Record, McGraw Hill Publishing Co. New York, N. Y., Mid August, 1976.

Cuttle, A. and Slater, A. I. "A Low Energy Approach to Office Lighting", Light and Lighting. Illuminating Engineering Society, 119 Westminister Bridge Road, London, England. Jan/Feb. 1975.

GSA, Energy Conservation Design Guidelines for New Office Buildings, Public Building Service, Government Services Administration. Washington, D. C., July, 1975.

6.3   AUTOMATIC SWITCHING/Daylighting

STRATEGY:

Control light switching with photo-electric light sensors or
timers to reduce electric lighting usage during periods of
adequate daylight.

PHENOMENA:

1)   Silicon cells generate a small electric current which
      varies directly with the light level. By measuring this
      current or measuring the change in resistance to a
      current passed through a silicon or selenium cell,
      changes in light levels can be sensed electrically.
      This information can be used by control systems to
      regulate electric lighting automatically according to
      need.

2)   Silicon cells are more stable and faster responding than
      selenium cells. Silicon cells are therefore better
      suited for controlling interior illumination. (Selenium
      cells are acceptable for less critical tasks such as
      switching security lighting.) (76,Zalewski)

3)   A timer can be built into the lighting circuit to limit
      the use of certain electric lights to night-time.

ADVANTAGES:

1)   Elimination of the situation of a cloud reducing the
      daylight level, the room occupant switching the lights
      on, and then leaving them on the remainder of the day.

2)   Reduced electricity demand for lighting and associated
      air conditioning.

3)   Elimination of night-time workers forgetting to turn out
      lights and daytime workers not noticing unneeded lights
      left on.

DISADVANTAGES:

1) Difficulty in purchasing "off the shelf" light-sensing systems. Only a few manufacturers are involved in their production. Furthermore, installation requires careful analysis on a case by case basis. A number of prototype systems were installed in the late 1950's and early 60's, but inexpensive energy greatly extended the amortization of first costs. The end of inexpensive energy now provides an incentive for installing light-sensing, light switching systems. Several manufacturers have already identified a market and are mass producing--or have in final design development--highly sophisticated, light-sensing switching systems.

2) Need for the sensors to be kept clean. This may require maintenance if they are mounted on the roof or outside walls.

3) Directional sensitivity of outdoor sensors. This requires careful analysis in predicting what the indoor illumination level will be for the outdoor light level at which the sensor responds by switching on or off indoor lights.

4) Greater variation in indoor light level when daylight is depended upon. Variation may be considered a disadvantage or an advantage depending on personal preference.

AESTHETICS:

1) The light sensor component of the control system can readily be made inconspicuous in the ceiling system or building facade.

2) Changes from side daylighting to overhead electric lighting during the course of a day provides a change in the character of the interior space. Interior forms are highlighted in a different manner and the direction and definitiveness of shadows change.

COSTS:

1) In addition to the hardware cost of the actual control
    system, consulting costs are incurred because lighting
    specialists must balance the system. Predicting day-
    light is a difficult problem because the absorption and
    reflection characteristics of both indoor and outdoor
    surfaces must be evaluated as well as the variability in
    the direction and intensity of the light source.

2) Circuiting the lights to additional switching controls
    is an added cost.

EXAMPLES:

1) A skylight manufacturer offers a daylight sensing, light
    control system. A time delay is built into the relay
    system so that a change in daylight level must last
    longer than ninety seconds. This reduces rapid cycling
    when clouds pass in front of the sun. A counter is also
    included to lock the lights on for a period of an hour
    if the lights are switched on and off more than three
    times in less than ten minutes. A clock can be included
    to provide automatic (as well as manual) over-ride to
    the system during weekends, holidays, evenings, lunch
    hours, and custodial hours. (76,Commendo)

2) A recent development is a solid state, dimming control
    system which varies the output of fluorescent or mercury
    vapor lighting according to the availability of daylight.
    When a silicon cell senses the daylight level, the
    control system varies the electric lighting output con-
    tinuously (rather than incrementally). (76,Longenderfer)

3) A photo activated interior venetian blind is manufactured
    in Great Britain. Banks of up to 30 blinds can be
    operated from one photo-cell unit. The control can be
    restricted to only tilting of the blind, raising or
    lowering being accomplished manually, or the entire
    operation of the blind system can be automated. (74,
    Beckett and Godfrey,p295)

4) An experimental system was installed in two classrooms
    of Heather Drive Elementary School in Aurora, Illinois.
    A sensor at the window turned off the row of lights

nearest the interior wall when the outdoor light level exceeded 1500 foot candles. A second row was turned off at 1000 foot candles, third at 850 foot candles, fourth at 700, fifth at 600, and the sixth (nearest windows) at 500. After carefully measuring the electric consumption for an entire semester, it was found that two unautomated classrooms (used as an experimental control) used 46 percent more electricity than did the classrooms with automated switching. (63,Chapman,p193)

5) New office space in a San Francisco building has been provided with ancillary lighting in certain areas which is controlled by a photo-cell override which limits its use to non-daylight hours only. (76,Architectural Record,p120)

6) One floor of the Manchester Federal Office Building is equipped with automatic switching. The three perimeter rows of lighting are connected to photo-electric cells adjacent to each row of lighting. When daylighting is adequate, each row is independently switched off. A 30 second time delay prevents the lights from rapidly switching on or off as would be the case with a cloud momentarily blocking the sun. (75,Isaak,p23)

REFERENCES:

Architectural Record, "Task Ambient Lighting", Architectural Record, McGraw-Hill Publishing Co., New York, N. Y., Mid August, 1976.

Beckett, H. E. and Godfrey, J. A., Windows, Performance, Design, and Installation, Van Nostrand Reinhold Co., New York, 1974.

Chapman, William P., "Automatic Controls Can Cut Lighting Costs", Architectural Record, McGraw-Hill Publishing Co., New York, May 1963.

Commendo, John, telephone conversation, Naturalite Skylights, Garland, Texas, June 1, 1976.

Isaak, Nicholas and Andrew, Designing an Energy Efficient Building, General Services Administration, Public Building Service, Washington, D. C., Sept. 1975.

Longenderfer, John, telephone conversation, Lutron Co., Allentown, Pa., June 9, 1976.

Zalewski, Edward, telephone conversation, Metrology Section, National Bureau of Standards, Washington, D. C., May 20, 1976.

6.4 INTERIOR COLORS/Daylighting

STRATEGY:

Paint interior surfaces a light color to increase the light
level possible from daylighting.

PHENOMENA:

1) Light colored surfaces reflect light increasing the
ambient light level for a given amount of available
light. Dark surfaces absorb light decreasing the
ambient light level. The following table gives typical
reflection factors for different colors: (75,Kern,p336)

REFLECTION FACTORS OF COLORS

COLOR	REFLECTION FACTOR
White	80 to 90 percent
Pale yellow, rose	80
Pale beige, lilac	70
Pale blue, green	70 to 75
Mustard yellow	35
Medium brown	25
Medium blue, green	20 to 30
Black	10

Note: Reflection factor = reflected light/incident
light given as a percentage.

The exact reflection factors for a given color can be
determined by matching it to a standard system of colors
called "Munsell Colors" and using a conversion table to
determine the reflectivity. (72,IES,p5-16)

2) The following are example reflection factors desirable
    for different surfaces of a room: (75,Kern,p336)

DESIRABLE REFLECTION FACTORS FOR INTERIOR SURFACES

SURFACE	REFLECTION FACTOR
Ceilings	80 percent
End walls:	
in poorly lighted rooms	70
in well lighted rooms	25
Walls containing window(s)	80
Floors	25

Note:

The use of light reflective colors on the wall(s) con-
taining window(s) decreases the contrast between the
windows and the surrounding surfaces. For the same
reason, the window frame, sash, and muntins should also
be a light color. Light colors on surfaces adjacent to
or opposite windows should have a matte finish to
alleviate the potential problem of reflected glare.

3) The likelihood of a room being adequately daylit is
    lessened if it is furnished with dark-colored draperies,
    carpeting, wall hangings, and furniture.

ADVANTAGES:

1) Increased availability of daylight for task illumination
    or ambient illumination.

2) Decreased likelihood of glare from excessive contrast
    between bright windows and dark surroundings.

DISADVANTAGES:

Dirt, fingerprints, and scuff marks are more conspicuous on
light-colored surfaces.

AESTHETICS:

Selected colors can have interesting effects upon people. A
"rule of thumb" is to use warm colors (yellows, oranges,
reds) in north-facing rooms receiving little or no sun, and
cool colors (greens, blues, and violets) in rooms receiving
plentiful sunlight.

COSTS:

The cost of light colored paints versus dark colored paints
is the same or insignificantly different.

EXAMPLES:

The following table illustrates the improvement in illumination
both in uniformity and brightness possible with increasing
reflectances of walls, floors, and ceilings. The illumination
at the rear of a room with very dark surfaces is used as the
basis of comparison. The room is 30 x 32 x 12 ft. high with
6 ft. directional glass block the full 32 ft. length. (66,
IES,p7-10)

EFFECT OF SURFACE REFLECTANCE UPON REL. ILLUMINATION

Reflectance Factor (percent)			Relative Illumination at various distances from fenestration		
Walls	Floor	Ceiling	3 feet	15 feet	17 feet
6	6	6	6.57	2.66	1.00
28	28	28	8.55	3.62	1.60
62	28	62	11.98	5.75	3.16

REFERENCES:

IES, IES Lighting Handbook, Illuminating Engineering Society, New
   York, N. Y., 1966.

Kern, Ken. The Owner Built House, Charles Scribner's Sons Inc.,
   New York, N. Y., 1975.

6.5   THERMAL MASS/Solar Heating

STRATEGY:

Locate massive materials directly in the path of winter sunlight
transmitted through windows in order to store part of the
incoming solar heat, avoid overheating, and provide re-radiated
heat during non-sunlit hours.

PHENOMENA:

1)   Large window areas, especially in small rooms, can transmit
      so much solar energy that overheating may occur during
      periods of peak solar intensity. This can be partially
      remedied with the use of heavy, massive materials. If the
      sunlight entering a room falls on the surface of a massive
      material, such as a slate floor, part of the energy is
      reflected as light, and part of the energy is absorbed and
      becomes heat. This heat raises the surface temperature of
      the material. When the internal temperature of the material
      is lower than the room air temperature, the heat is con-
      ducted inward into the mass of the material. When the room
      temperature later begins to drop and becomes lower than the
      surface temperature of the material, the heat accumulated
      within the material flows outward. The heat is returned to
      the room by convection at the surface of the material, and
      by radiation from the surface of the material to opposite
      and adjacent room surfaces with a lower temperature.

2)   How effective a material is in storing heat can be judged
      from its ability to absorb sunlight, conduct surface heat
      into its mass (conductivity), and hold the resulting heat.

      The ability of a material to absorb sunlight is largely
      determined by its color and texture. The following
      table provides an approximation of the percent of solar
      radiation absorbed by different colors. (65,ASHRAE,p1-
      6)

COLOR	PERCENT ABSORPTION
white, smooth surfaces	25 to 40 percent
gray to dark gray	40 to 50
green, red, brown	50 to 70
dark brown, blue	70 to 80
dark blue, black	80 to 90

3) The ability of a material to hold heat can be judged by its
    "thermal capacity" which equals its density multiplied by
    its specific heat. (Density: Pounds per cubic foot or
    kilograms per cubic meter; Specific heat: BTU/lb°F or
    KJ/Kg°C.)

    The greater the thermal capacity, the better the material
    is for storing heat. If two materials have a similar
    thermal capacity, the material with the higher conductiv-
    ity is a better storage medium. The following table
    provides information on both solar absorption and thermal
    capacity of several common materials. (Calculated from 67,
    Baumeister, p4-11 and 6-8)

SUITABILITY OF MATERIALS FOR STORING HEAT FROM SUNLIGHT

MATERIAL-COLOR	TOTAL PERCENT SOLAR ABSORBED[1]	THERMAL CAPACITY (BTU/°F,ft.[3])	CONDUC- TIVITY BTU/hr ft°F
Brick - glazed white	0.26	24.6	0.4
Brick - common, red	0.68	24.6	
Marble - white	0.44	35.7	1.5
Marble - dark	0.66	35.7	
Granite - reddish	0.55	32.2	
Slate - blue gray	0.87	37.8	
Slate - dark gray, rough	0.90	37.8	
Concrete -	0.65	22.5	0.54
Wood - white pine		18.1	0.06
Wood - white oak		27.4	0.10
Steel - enamel red	0.81	58.8	26.2
Water		62.0	0.35

NOTES:

    1. Source: (65,ASHRAE,p1-61)

    2. Conductivities are given without regard to color in
        72,ASHRAE,p570.

4) The following table can be used as a guideline to roughly
    approximate how much mass should be provided to store a
    window's solar heat gain. (Square feet of material per

square foot of window.) Different areas of material are given for different thickness of materials, for different amounts of daily solar heat gain, and for different average inside-outside temperature differences. (74, Anderson, p46) The inside-outside temperature difference is important because it determines how much heat is needed to maintain a comfortable room temperature. Not given is the resistance to outward heat flow afforded by the building envelope. A well insulated double glazed residence would require less mass.

AREA OF CONCRETE REQUIRED TO STORE WINDOW SOLAR HEAT GAIN (SQ.FT.)

Average Daily Solar Heat Gain Through the Window[1]	Allow. Inside Temp. Swing (°F)	THICKNESS OF CONCRETE[2]			
		2"	4"	8"	12"
500 Btu/ft^2	5	25.00	12.50	6.25	4.17
	10	12.50	6.25	3.13	2.08
	15	8.34	4.17	2.08	1.39
	20	6.25	3.13	1.57	1.04
	25	5.00	2.50	1.25	0.83
750 Btu/ft^2	5	37.50	18.75	9.38	6.25
	10	18.75	9.38	4.69	3.13
	15	12.50	6.25	3.13	2.08
	20	9.38	4.69	2.35	1.56
	25	7.50	3.75	1.87	1.25
1000 Btu/ft^2	5	50.00	25.00	12.50	8.33
	10	25.00	12.50	6.25	4.17
	15	16.67	8.34	4.17	2.78
	20	12.50	6.25	3.13	2.08
	25	10.00	5.00	2.50	1.67
1250 Btu/ft^2	5	62.52	31.25	15.63	10.42
	10	31.25	15.63	7.81	5.21
	15	20.83	10.42	5.21	3.47
	20	15.63	7.81	3.92	2.60
	25	12.50	6.25	3.13	2.08
1500 Btu/ft^2	5	75.00	37.50	18.75	12.50
	10	37.50	18.75	9.37	6.25
	15	25.00	12.50	6.25	4.17
	20	18.75	9.37	4.68	3.12
	25	15.00	7.50	3.75	2.50
1750 Btu/ft^2	5	62.52	31.26	15.63	10.42
	10	31.26	15.63	7.81	5.21
	15	20.83	10.42	5.21	3.47
	20	15.63	7.81	3.92	2.60
	25	12.50	6.25	3.13	2.08

Notes:

1) The average daily solar heat gain for a specific region and orientation can be found on pages 388 through 392 of the 1972 edition of the ASHRAE Handbook of Fundamentals.

2) To estimate the area of other monolithic materials required, multiply the areas given in the table by the following values:

wood   1.00   stone   0.83
brick   0.86   water   0.38

5) Massive materials may be effective during the summer, if the night temperatures are substantially lower than into the daytime temperature. Outdoor night air allowed to enter through open windows and passing over a massive material will lower its surface temperature. Subsequently, the internal heat will be drawn out of the material. During the day, if the outside air temperature is greater than the indoor air temperature, the windows are closed and shaded. As the room air temperature rises, heat is absorbed back into the cooler massive material. The rate at which the room air temperature rises is thereby reduced.

6) The geometry of a building material will also affect its ability to store winter solar heat or summer evening coolness. Steel is a good storage material due to its high density (in spite of its low specific heat). However, if the material occurs in a shape with a great deal of surface area such as a wide flange beam, there is very little thickness where the heat can be stored unless for a short period of time. The large surface-area to volume ratio limits the ability of the beam to store heat.

7) The effectiveness of a massive material in augmenting the heating or cooling system can be increased by embedding air ducts or copper tubing in the material and routing the heat transport medium of the mechanical system through the slab.

ADVANTAGES:

1) Increased utility of solar energy transmitted through windows

2) Delayed day time temperature rise during the summer.

3) Reduced frequency of heating or cooling system cycling (switching on and off) due to the "thermal inertia" of the massive material.

4) Reduced sound transmission when massive materials are used between spaces to be acoustically separated.

DISADVANTAGES:

1) Decreased responsiveness of room temperatures to the heating or cooling system.

2) Increased reverberation of sounds within the room due to the surface hardness common to dense, massive materials.

3) Decreased flexibility in furniture arrangement.

4) Decreased flexibility in repartitioning floor space when massive materials are used in the interior wall construction.

5) Increased loading which must be carried by the structural system. (However, the massive material may be part of the structural system as with load bearing interior masonry walls.)

AESTHETICS:

1) Massive walls, floors, or ceilings give a sense of permanence.

2) The hard surfaces of massive materials may be smooth or textured to reduce their harshness.

3) Tapestries, rugs, and other coverings must be located so as not to reduce the effectiveness of massive materials.

COSTS:

The cost of massive materials such as concrete, marble, slate, or brick flooring may be justified on an aesthetic, durability, or structural basis with the thermal mass considered a supplemental benefit.

EXAMPLES:

1) The National Concrete Masonry Association, using the National Bureau of Standards Computer Program (NBSLD) to evaluate the effects of mass on the size of air

conditioning equipment, calculated a 16 percent reduction
in peak cooling load could be achieved through the use
of massive materials. In the first series of computer
runs a 2-story building was exposed to a 24-hour-cycle
of temperatures ranging from 76°F to 94°F. For the
first computer run the walls were assumed to be composed
of wood studs and insulation having a weight of 8 lbs.
per square foot and U-value of 0.10. Results of the
analysis indicated that maintaining 75°F would require
air conditioning equipment with a peak cooling load
capacity of 40,500 Btu per hour. For the second
computer run the walls were changed from insulated
wood frame to insulated concrete masonry with the same
U-value, but the weight was increased to 40 lbs. per
square foot (8" light-weight concrete block). The
peak load was then calculated to be 34,000 Btu, or a
savings of 6,500 Btu, a reduction of 16 percent.
(75,NCMA-TEK,p2)

2)   A house in Santa Fe, New Mexico is entirely heated
      from solar heat transmitted through 384 sq. ft. of
      south-facing double glazed windows. To absorb the
      heat and keep the room temperature comfortable, the
      house has a large amount of interior mass. Walls are
      14-inch adobe, floors are brick with 24 inches of
      underlying adobe, and several benches in the path of
      the incoming sunlight contain 55 gallon drums of
      water. This mass is capable of keeping the home
      comfortable for three to four sunless days. (76,Cole,p22)

3)   A vacation house in Illinois built by the owners for
      approximately $10,000 is capable of achieving tempera-
      tures up to 100 degrees with window solar heating
      (37.8°C) when the outside temperature is five degrees
      (-15°C). One of the key components of the system is
      the provision of heat storage. Two 16 cubic feet
      insulated steel tanks filled with water are located in
      the path of sunlight entering through side windows and
      skylights. Opened reflective panels outside the
      window reflect additional sunlight through the skylights
      during the heat collection period, then the panels are
      closed at night to help reflect the heat radiating
      from the storage tanks back into the room. In the
      summer the process is reversed: the panels are closed
      during the day then opened at night to allow the heat
      accumulated in the water from the room to be dissipated
      to the night sky. (75,House and Garden,p134)

REFERENCES:

Anderson, Bruce, et.al. Solar Energy House Design in Four Climates, Total Environmental Action, Harrisville, N. H., May 1975.

Balcomb, J. D., Hedstrom, J. C. Simulation Analysis of Passive Solar Heated Buildings, LA-UR-76-89, Los Alamos Scientific Laboratory, Los Alamos, N. M., 1976.

Baumeister, Theodore (Ed.) Standard Handbook for Mechanical Engineers, McGraw Hill, New York, N. Y., 1967.

"Energy and Buildings - Thermal Analysis: Part 2" Architects Journal, 9 Queen Anne's Gate, London, England, March 24, 1976.

Loudon, A. G., "Window Design Criteria to Avoid Overheating by Excessive Solar Heat Gains", Building Research Station; Garston, Watford, England, April 1968.

NCMA-TEK, "New Findings on Energy Conservation with Concrete Masonry", National Concrete Masonry Assoc., McLean, Va., 1975.

PCA, "The Concrete Approach to Energy Conservation", Portland Cement Assoc., Skokie, Ill., 1974.

Total Environmental Action, Solar Energy Home Design in Four Climates. Total Environmental Action, Harrisville, N. H., 1975.

"Vacation House Warmed by Sun Power", House and Garden, Conde Nast Pub., New York, N. Y., 1975.

CONCLUSION

A window can be a solar collector introducing valuable
energy which can lower winter heating costs; a source of illumination which can substitute for artificial lighting to lower
electricity expenditure; and a means of natural ventilation which
can postpone the need for air conditioning in the spring and
fall, and substitute for air conditioning on cool summer evenings.

Numerous design strategies can improve these capabilities of
a window. The solar energy a window receives can be increased by
light-colored adjacent ground surfaces and by favoring southern
exposures. The usefulness of sunlight inside the building can be
increased by providing mass to store part of the sun's heat. The
utility of daylight can be increased by providing light-colored
walls and ceilings, and by facilitating the substitution of
daylight for electric light. Examples include separate switching
of perimeter lighting, task lighting separate from ambient
lighting, and automatic control systems driven by light sensors
and/or timers. Finally, the ability of windows to provide ventilation can be improved through proper orientation to prevailing
winds and by selecting operating window types which effectively
direct the entering and exiting air stream. Even when fixed
glass is required opportunities for admitting outside air are
available with frame ventilators or thru-glass ventilators.

Design strategies can likewise minimize the window thermal
load on mechanical systems. Winter heat loss through windows can
be reduced with double glazing, storm sash, or edge-sealed transparent roll shades. Night-time heat loss can be minimized with
tight-fitting draperies, opaque roll shades, or insulating shutters
Leakage of unconditioned outside air in and conditioned inside
air out through window cracks can be greatly reduced by initially
installing good quality windows, by providing weatherstripping,
and by landscaping and exterior appendages which reduce the force
of the wind. Finally, there are numerous means of blocking solar
heat gain in the summer, the most effective solutions being
exterior appendages or site solutions.

Window design strategies can provide occupants with more
freedom in managing their individual environments, and when
effectively used, they can improve comfort, and reduce purchased
energy expenditures.

ACKNOWLEDGMENTS

N.B.S. Reviewers:	Belinda Collins
	Porter Driscoll
	Tamami Kusuda
	Donald Quigley
	Nancy Starnes
	Heinz Trechsel
	Robert Wehrli
Bibliographic Assistance:	Elsie Cerutti
	Elizabeth Warner
Graphic Assistance:	Nan Stephens
	Jerry McKay
Typing:	Cheryl Talley

www.ingramcontent.com/pod-product-compliance
Lightning Source LLC
Chambersburg PA
CBHW051211200326
41519CB00025B/7072

* 9 7 8 1 4 1 0 1 0 8 6 3 0 *